Synthesis Lectures on Information Concepts, Retrieval, and Services

Series Editor

Gary Marchionini, School of Information and Library Science, The University of North Carolina at Chapel Hill, Chapel Hill, NC, USA

This series publishes short books on topics pertaining to information science and applications of technology to information discovery, production, distribution, and management. Potential topics include: data models, indexing theory and algorithms, classification, information architecture, information economics, privacy and identity, scholarly communication, bibliometrics and webometrics, personal information management, human information behavior, digital libraries, archives and preservation, cultural informatics, information retrieval evaluation, data fusion, relevance feedback, recommendation systems, question answering, natural language processing for retrieval, text summarization, multimedia retrieval, multilingual retrieval, and exploratory search.

Kathleen Gregory · Laura Koesten

Human-Centered Data Discovery

 Springer

Kathleen Gregory
Faculty of Computer Science
University of Vienna
Vienna, Austria

Laura Koesten
Faculty of Computer Science
University of Vienna
Vienna, Austria

School of Information Studies
Scholarly Communications Lab
University of Ottawa
Ottawa, Canada

ISSN 1947-945X   ISSN 1947-9468 (electronic)
Synthesis Lectures on Information Concepts, Retrieval, and Services
ISBN 978-3-031-18225-9   ISBN 978-3-031-18223-5 (eBook)
https://doi.org/10.1007/978-3-031-18223-5

This Springer imprint is published by the registered company Springer Nature Switzerland AG
The registered company address is: Gewerbestrasse 11, 6330 Cham, Switzerland

Acknowledgements

We thank all our collaborators, as well as past research participants.

Parts of this work were based on the authors' Ph.D. work. We therefore thank our advisors: Sally Wyatt, Paul Groth, Andrea Scharnhorst, Elena Simperl and Jeni Tennison.

We also recognize the funding we received during our doctoral work from the following grants: European Union's Horizon 2020 research and innovation program under Marie Skłodowska-Curie grant No 642795, Data Stories project, funded by EPSRC research grant No. EP/P025676/1; the NWO Grant 652.001.002 Research: Contextual Search for Scientific Research Data.

Contents

Introduction

1

1.1   Data, Data Everywhere … But Can We Find What We Need?

Data are everywhere. Across sectors, millions of datasets are available in data repositories, online marketplaces and from individual publishers (Brickley et al. 2019; Verhulst and Young 2016). By 2013, for example, one million datasets were already available in open governmental data portals worldwide (Cattaneo et al. 2015); by August 2020, the number of indexed datasets in the Google Dataset Search corpus had reached 31 million (Noy and Benjelloun 2020), which is still likely to be just a snapshot of the data currently available on the web (Benjelloun et al. 2020).

The increased amount of available open governmental and research data is in part due to new policies and guidelines for open science, research data management, and open data at local, national and international levels (Zuiderwijk and Janssen 2014), some of which are supported by major research funders (e.g. Directorate-General for Research and Innovation 2021; National Institutes of Health 2003).

Such policies are situated in claims about the potential benefits of reusing available data, claims which are also echoed within business and financial sectors. Tapping into vast digital data supplies is seen as a way to answer old and new questions (Hey et al. 2009), to glean new business insights (Gartner 2022), to reduce costs (European Commission 2018; European Commission 2011), and to increase transparency and encourage civic participation (McDermott 2010). This emphasis is also visible in the increased demand for, and spending on, financial, economic, and marketing data provided by vendors such as Bloomberg or Thomson Reuters (Burton-Taylor 2015), as well as through the sharing of large datasets internally within organizations (Halevy et al. 2016).

Despite the increase in data supplies, the attention paid to their potential benefits, and the extensive body of research investigating information-seeking behaviors (Case and

Given 2016), we still know relatively little about how people *find data*, how they *understand data*, and how they put *data to use* (Brickley et al. 2019). This lecture consolidates existing work on different aspects of *human-centered data discovery* with the aim of stimulating future work in the area and informing human-centered solutions for discovering data.

1.2   Data Discoverability, Discovering Data

1.2.1   Making Data Discoverable: Technical Challenges

Data are collected, curated and shared in various ways. Large amounts of sensor data, for example, are stored, processed and curated through a variety of automated as well as manual workflows (Kitchin and McArdle 2016). Data produced through academic research, including big data stemming from large-scale projects and smaller data collected in short-term projects (Borgman 2012, 2015), are deposited and curated in data repositories.

Sharing data does not ensure that they are discoverable or usable. As one step to address this, work is being undertaken to make data *findable, accessible, interoperable and reusable* (FAIR) through the use of machine readable standards (Wilkinson et al. 2016). Projects and organizations assign persistent identifiers to data and connect them with other digital research objects (e.g. DataCite[1]). Others develop metadata standards or knowledge organization systems which enable discoverability and interoperability between siloed data, which are distributed across different locations on the web (Sansone et al. 2019).

The emerging area of *data search* within the information retrieval community (see Koesten et al. 2019a) has also focused on creating new ontologies and standards for data (Ohno-Machado et al. 2017; Sansone et al. 2017). Researchers here have explored methods to semantically enhance datasets (Khan et al. 2016), sought ways to apply traditional keyword searching to datasets (Lopez-Veyna et al. 2012) and thought about tabular search, where tables serve as queries (Zhang and Balog 2018). While much of data search has evolved heavily based on what is known about web search, a key difference is the almost complete reliance on metadata within data search.

Complete, well-curated metadata are critical to making data discoverable. Data search systems and search engines, including Google Dataset Search,[2] rely on metadata to search data, as techniques which work well for other types of search (i.e. document search) do not work equally well for searching data (Chapman et al. 2020; Cafarella et al. 2011). Whether performed via global web crawls (Brickley et al. 2019; Sansone et al. 2017) or within organizational holdings (Hendler et al. 2012; Kassen 2013), keyword-based searching for data is executed by searching the metadata published alongside data, rather

---

[1] https://datacite.org/.

[2] https://datasetsearch.research.google.com/.

than by searching through the data themselves. In practice this means that even if a dataset contains a specific keyword as a variable, i.e. in the name of a column header in tabular data, the dataset may not be retrieved if the keyword is not explicitly mentioned in the associated metadata.

Metadata vary, however, in their availability, granularity and quality (Atz 2014; Brickley et al. 2019). Even in disciplinary domains where established documentation standards and metadata schemas exist, their creation practices differ and standards may not be followed in the same way by everyone (Edwards et al. 2011). The resulting metadata describing data are often incomplete (Robinson-Garcia 2017), which hampers both reuse and discoverability.

Existing algorithms for web search engines designed to rank web pages (Page et al. 1998) also do not equally support the indexing of structured content such as data. While some aspects of data discovery are quite advanced, i.e. search within databases (Wang and Zhang 2005), other aspects, such as techniques for integrating heterogeneous data scattered across the web, remain in development. Information retrieval technologies designed primarily to work on unstructured documents are also less tailored to meet the specific needs which people have when looking for data (Wilson et al. 2010; Cafarella et al. 2008); and efforts to identify effective ways to rank datasets for particular tasks are hindered by an overall lack of knowledge about how people search for data (Brickley et al. 2019).

1.2.2   Discovering Data

Much of the existing work in the area of data search has focused on *making data discoverable*; substantially less work has investigated the practices involved in discovering data, or *human-centered data discovery*, to explore how people find, evaluate and decide to use data to meet their various data needs.

Discovering data is not like searching for literature, which has been well-studied within information and computer science and is well-supported (Kern and Mathiak 2015). Data discovery requires extra considerations for various actors, such as taking into account issues of data provenance, granularity of content and diverse access restrictions (Chapman et al. 2020). Data needs range in specificity, and are closely related to an individual's tasks and purposes (Koesten et al. 2017), although there is also an overlap between the sources and strategies used to discover data and those used to discover other information, i.e. academic literature (Gregory et al. 2020a). Understanding data needs and discovery practices, as well as their relation to other information behaviors and infrastructures, is key to developing sustainable solutions for discovering, understanding and using data.

Current work examining *data discovery from a user perspective* is scattered throughout the literature. Existing studies center on different types of data, i.e. open government data or research data (Kacprzak et al. 2019; Gregory et al. 2020b); different systems and infrastructures for making data available, i.e. within specific disciplinary data repositories or more general data portals (Martín-Mora et al. 2020); or different types of users. i.e. researchers, data scientists, data journalists or civil servants (Koesten et al. 2017). Insights about data discovery practices are often buried within studies of other data practices, such as data sharing, data management or data reuse.

A range of methodologies, rooted in different areas of computer and information research, have been used to explore human-centered data discovery. Search logs (Kacprzak et al. 2019), analyzing question corpora (Löffler et al. 2021; Roberts et al. 2017; Kato et al. 2020), surveys (Ariño et al. 2013; Gregory et al. 2020a), usability evaluations of data search systems (Dixit et al. 2018), literature studies (Gregory et al. 2019), in-depth interviews (Koesten et al. 2017; Krämer et al. 2021), and ethnographic ⸱pproaches (Borgman et al. 2019), have all been utilized to explore data needs and discovery practices of different communities. Many studies propose recommendations for improving data discovery, although these recommendations, like data themselves, often remain strangely siloed, high-level, and have yet to be brought together in one place.

1.3   Aim and Structure of the Lecture

This lecture synthesizes existing work on *human-centered data discovery* from across traditions, foci, and methodologies, including recommendations for the design of data discovery solutions and guidelines to support finding, evaluating and understanding data for reuse.

The lecture is divided into six chapters, including this introductory chapter. Chapter 2 defines common terms used throughout the lecture and situates *human-centered data discovery* within the broader field of information-seeking research, which has been well explored in computer and information science. We then synthesize and present existing work according to the *data needs* of individuals and communities in various contexts in Chap. 3, before proceeding to examine common data sources and discovery strategies in Chap. 4. The fifth chapter focuses on evaluation criteria and practices of data-centric sensemaking. We conclude by consolidating and proposing recommendations for data discovery systems and publishing practices in view of user needs in Chap. 6, and at the same time consider future directions for research in the field of human-centered data discovery.

1.4   Definitions for Easy Reference

As with information (Floridi 2010), "data" is a term that has been variously defined, and one that continues to evolve, as evidenced by the long-running debate about the use of the singular or plural to refer to data. In this work, we refer to data in the plural, reflecting the fact that data are not isolated entities, but are always rooted in context and related to other practices, technologies and documentation (see Leonelli 2015).

Below, we present a brief overview for easy reference of the data-related terms (Table 1.1) which we use throughout this manuscript, as well as the practice-related terms relevant to data discovery (Table 1.2). The final column in each table provides references to the work in which these definitions have their basis. We discuss these terms in more detail in Chap. 2.

Table 1.1 Data-related terms and definitions

Term	Definition	References
Data	We understand data broadly, as there is a great range and diversity in what serves as data Data can include geospatial coordinates, numerical values and measurements, but also literature corpora, images or physical samples—**any of which may be used to provide evidence of a phenomena or to serve as a subject of analysis** Data can be organized in different ways, i.e. in spreadsheets, as networks or graphs, or as collections of related artifacts	Informed by Borgman (2015), Dourish and Gomez Cruz (2018), Munzner (2014)
Structured data	Data that are explicitly organized and formatted, i.e. in spreadsheets, web tables or relational databases, usually for a particular purpose	Losee (2006), Koesten (2019)
Open governmental data	Data which are provided to the public by a governmental agency, usually under an open license This term is also used to refer "public sector data"	Zuiderwijk et al. (2015)
Research data	Entities used for the purpose of evidence in academic work; term also used for *scholarly data* or *scientific data* Data are relational in that different entities will serve as data in different communities and at different timepoints	Borgman (2015), Leonelli (2015)
Dataset	A collection of data including primary data and metadata organized and formatted for a particular purpose	Chapman et al. (2020), Löffler et al. (2021)

(continued)

Table 1.1 (continued)

Term	Definition	References
Metadata	Informally referred to as *data about data* More specifically, metadata are the documentation and other data necessary for the identification, representation, interoperability, technical management, performance, and use of data. Textual description of datasets can be seen as part of metadata	Informed by Gilliland (2008), Pomerantz (2015)
Data needs	Data which individuals desire or require for a particular purpose Data needs may be implicit or explicit	Informed by Taylor (1968)

Table 1.2 Practice-related terms and definitions

Data search	The process of individuals actively engaging with search systems to locate data; the search process, the ranking and the return of datasets	Informed by Wilson (1999), Chapman et al. (2020)
Data seeking	A broader set of goal-directed practices for finding data, potentially involving a variety of resources, not only interacting with search systems	Informed by Wilson (1999)
Data discovery	The umbrella term for the entire process of acting on a data need This can include data search and seeking, but it also allows for serendipity to occur and is not limited to goal-directed behaviors	Informed by Wilson (1999)
Data-centric sensemaking	The process of understanding and interpreting data, informed by context	Koesten et al. (2021)
Data reuse	Using data which others have created, for any purpose	Informed by Fear (2013), van de Sandt (2019)

Data Discovery: A Human-Centered View

2

This chapter (i) contextualizes the terminology used in the remainder of the manuscript and (ii) situates human-centered data discovery within the context of existing information-seeking research. We end the chapter by providing examples of data seeking in practice, drawn from our research, to further contextualize human-centered data discovery.

2.1   What Do We Mean by Data?

There are many diverse definitions of data. Even in studies about people's perceptions of data, the range of information objects that are considered to be data is wide. The concept of data is also abstract, with definitions varying depending on the particular community (Borgman 2012; Pasquetto et al. 2017). The line between data and information is nearly impossible to draw (Pasquetto et al. 2017), and making a clear distinction between the two might not always be useful. A large body of work discusses the nature of data and its relation to practice and reuse, emphasizing how different communities of practice see and use data differently (Borgman 2015; Wynholds et al. 2012).

In the previous chapter, we formulated a broad definition of data characterized by how data are used, i.e. as evidence of a phenomena or as subjects of analysis. While various objects serve as data (see Table 1.1), data are often published in a dataset, or in multiple datasets, which are organized for a particular purpose. This is in line with Chapman et al.'s definition of datasets *as a set of related observations, organized for a particular purpose and then released for consumption and reuse* (2020).

Common data formats for organized, or structured, data include tabular formats, CSVs or JSON. Other structured data types also include graph, network data or geospatial data, and, depending on the working definition of data, can also include simple textual lists, collections of images or even physical objects.

© The Author(s), under exclusive license to Springer Nature Switzerland AG 2022
K. Gregory and L. Koesten, *Human-Centered Data Discovery*, Synthesis Lectures on Information Concepts, Retrieval, and Services,
https://doi.org/10.1007/978-3-031-18223-5_2

In this lecture, we synthesize literature from across different disciplines. As we do so, we wish to highlight that every work we discuss likely has its own definition of data which needs to be taken into account when thinking about the generalisability and comparability of findings or frameworks.

2.1.1 Data as an Object Versus Data-as-Process

Data (or information) can be seen as something material, **as an object**, that can be stored, found, accessed, retrieved, consumed, maintained, archived, sent somewhere, looked at or sold. In this sense data can be a set of images, graphs, or documents, as well as the classically conceived spreadsheet. Data are often stored in specific data repositories, such as for research data, e.g., Figshare (Thelwall and Kousha 2016) or Dryad,[1] or in open data portals[2,3] (Hendler et al. 2012; Kassen 2013). These data can be discovered by search engines embedded within repositories or by more general search engines which work across indexed repositories, i.e. Mendeley Data.[4] Data-specific search engines which attempt to query the whole web, such as Google Dataset Search (Brickley et al. 2019), are a relatively recent development.

Thinking about data as something that can be queried for and retrieved by a search system is a common perspective in the field of information retrieval. This discipline tends to take an "information-as-thing" perspective where information is viewed as a physical object which can be acted upon (Buckland 1991). Data are thus seen as something which can be retrieved and returned to a data seeker after a search activity for further evaluation and exploration. The more structured the data are, the more questions arise about whether search systems can, or should, take advantage of such structure.

On the other hand, other work takes a **data-as-process** perspective, similar to Buckland's description of "information-as-process" (Buckland 1991). In this view, information, or data, are situational and have multiple definitions which might hold only in relation to *context*. Borgman, talking about research data, describes that what count as data often lies in the eye of the beholder (2015), not least because of the different philosophical stances of particular epistemic communities (Borgman et al. 2007; Wallis et al. 2007). Leonelli, further points to the influence that different timepoints in the research process can have on the notions of data (2015). Data are determined in context, at specific times, according to a purpose which tells us something about how data can be used but also why they were originally created. While data cannot be fully defined without context, we also need context to understand data.

---

[1] https://datadryad.org/.

[2] https://ckan.org/.

[3] https://data.gov.uk/.

[4] https://data.mendeley.com/.

2.1.2   Context

Context is again a complex and variously defined concept. In 1997, Dervin proposed three approaches to conceptualize context which still resonate today (Dervin 1997). According to Dervin, information researchers tend to view context as a backdrop of a phenomena, where context is seen to be a setting or a container of an event or an interaction which includes virtually anything except the phenomena of interest. Alternatively, information researchers think of context as a list of stable elements which can be enumerated and named and which affect an interaction or event. The final view of context which Dervin identifies is to see context as the "inextricable surround without which any possible understanding of human behavior becomes impossible" (p. 113). In this view, context is portrayed as a set of relationships between actors, objects and activities that is enacted and local to each activity (Dourish 2004). As Dourish summarizes, "context isn't described as a setting; it is something that people do" (Dourish 2004, p. 22).

In a synthesis lecture on context, Agarwal extensively reviews the topic to arrive at a unifying definition of context. He identifies 12 themes across the field and ultimately describes context as consisting of elements "*such as environment, task, actor-source relationship, time, etc. that are relevant to the behavior during the course of interaction and [that] vary based on magnitude, dynamism, patterns and combinations…that appear differently to the actor than to others, who make an in-group/out-group differentiation of these elements…*" (Agarwal 2017).

While a clear definition of context as it pertains to data remains challenging (Faniel et al. 2019), there is general consensus that data reuse without any contextual reference is in most cases impossible (Birnholtz and Bietz 2003; Borgman et al. 2015). Some data are more self explanatory than others; however, additional information is almost always needed in order to "place" data into different contexts (Koesten et al. 2021). We therefore take wider social and disciplinary contexts around data work into account when considering data discovery (Chap. 5) and emphasize the impact of context on people's engagement with data.

In a search setting, selecting data is influenced by the information exposed with data, the interface design, and the different norms around certain data, including how data are described, communicated and published. Contextual information about data can be explicitly provided as metadata; other information is implicitly available via contextual elements, such as the publishing institution or repository and its reputation, or indications of prior data use. Such information can be used to aid result selection and to provide clues about a dataset's context to a data seeker in an online environment.

Metadata are not the only carriers of explicit contextual information. Such information can also take different forms and be carried in different *information structures*, discussed further in Chap. 5, which include codebooks, journal articles, email exchanges, or annotations within a spreadsheet or on an image.

2.2   What Do We Mean by Data Discovery?

2.2.1   Data Discovery in Relation to Information-Seeking Models

Information-seeking research, which examines how people discover information in order to meet a particular need, has been a topic of significant interest within information science and computer science (Courtright 2007; Marchionini 1997).

Information behavior (IB) and information retrieval (IR) research are areas of information science and computer science that are closely-related, yet separate, with different areas of emphasis (Stock and Stock 2013). Information retrieval, rooted in computer science, focuses on technologies that support finding and presenting of information; the field was traditionally concerned with developing algorithms that improve precision and recall in search processes (Sanderson and Croft 2012; Courtright 2007). There is a rich body of literature within IR which explores how people select documents and determine their relevance to a given task or information need. The field has produced a number of user-centered models to explain how people search for, evaluate and use information (Case and Given 2016).

Information behavior on the other hand puts the emphasis on people's information needs and how they seek and use information (Courtright 2007) and is rooted more within information science. A range of activities is included in the IB perspective, including the accidental encounter of information, in addition to needing, finding, choosing and using information (Agarwal 2015). Rooted in this perspective, numerous user-centered information behavior models have also been developed, many in the last decades of the twentieth century, which continue to motivate new research and remain relevant for systems design (White 2016; Marchionini and White 2009), although they do not generally mention data explicitly.

Models from both the IR and IB perspectives usually describe relationships among concepts related to information-seeking activities (Case and Given 2016). While conflicting views on the applicability of such models exist, they are pervasive in computer and information science, and are used not only to frame research into search practices but also to guide the design and development of new search interfaces (Marchionini and Komlodi 1998; Marchionini and White 2009).

Some models provide a high-level summary of the activities involved in finding and using information (Xie 2008), others are more stage-like and analytical in their design (Savolainen 2018), representing processes and strategies users take when searching for information (White 2016). We give a very high level overview of some of the key information-seeking models, with no claim to comprehensiveness, in the following subsection; for detailed reviews, refer to e.g. White (2016), Xie (2008).

2.2.1.1 Review of Key Information-Seeking Models

Ingwersen's cognitive model of information transfer refers to cognitive transformations that occur between five elements involved in information-seeking: a user's cognitive space, their socio-organizational environment, the interface/intermediary, the information objects, and the search system (1992, 1996). The model considers cognitive structures or knowledge structures that represent a model of the world. These structures are dynamic and change corresponding to a user's social or collective experiences (Ingwersen 1992).

Saracevic's stratified model of information retrieval interaction depicts interaction in information-seeking as an interplay between different levels representing users and systems. The dynamic nature of interaction is at the heart of this model, where search is portrayed as a dynamic conversation between a user and a search system. Both the user and the system contribute to this conversation, the user by entering search queries, browsing and navigating, and the system by providing results or suggestions (Saracevic 1996, 1997).

Belkin's episodic model of interaction with texts (1996) describes information-seeking along different dimensions, together with associated information-seeking strategies identified in his earlier work on interaction with text (1993). This model emphasizes interactions by including the user as an inherent part of the system (Xie 2008), and recognizes that a searcher will employ a variety of different search strategies over multiple search episodes.

Ellis's model for information-seeking behavior applies a more behavioral approach. This model describes six main characteristics of the information-seeking process, namely *starting, chaining, browsing, differentiating, monitoring, and extracting* (Ellis 1989; Ellis et al. 1993). The model, which integrates a dynamic perspective of the information-seeking process, was used to investigate information-seeking patterns of different user groups. The model has been cited widely in information-seeking and retrieval research and has contributed to a human-centered focus within information research (Xie 2008).

A more recent model that has been taken up in studies of data discovery is the *information journey model*, proposed by Blandford and colleagues (Adams and Blandford 2005; Blandford and Attfield 2010). The model reflects many aspects from different information-seeking models, while including notions of sensemaking and use. The model describes different phases: from recognising an information need to acquiring, interpreting and validating the information, to finally using this interpretation. Importantly, these phases are not necessarily seen as sequential. Marchionini and White (2007) describe a very similar set of activities, but focus more explicitly on the examination of search results.

2.2.1.2 Applying Information-Seeking Models to Data Discovery

Most of these models are either based on textual concepts of information or they do not actively consider data as the information object to be discovered. More recent work, including our own, has investigated information-seeking explicitly for data in the context of open governmental data, research data or data on the web by different user groups (Gregory 2021; Koesten 2019; Kern and Mathiak 2015; Friedrich 2020). Much of this

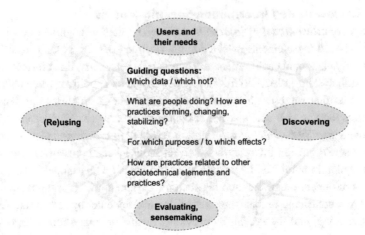

Fig. 2.1 Gregory's integrated theoretical framework of data discovery in context. Adapted from Gregory (2021)

work has been conducted relatively recently by early career researchers and has been published within doctoral dissertations.

Gregory proposes a modified version of the information journey model by combining Information Science and Science and Technology Studies perspectives to provide a framework for studying practices of data discovery which take context into account (Gregory 2021; Fig. 2.1). This integrated framework was used to guide empirical work with researchers and describes user needs in relation to data discovery, evaluation of data, and use of data. Context here is described as a web of sociotechnical elements, including, but not limited to, data, search systems, communities and other practices of (re)search.

The dynamic and enacted nature of context is acknowledged in this framework. Users are seen as social actors, situated in different types of communities who engage with technologies in various ways. Gregory argues that data discovery practices need to be understood as part of their communities, taking into account the types of data and the technologies used within particular contexts. The framework proposes a list of guiding questions to inform research into data discovery. These questions encourage looking for communities and data which are not represented; focusing on actual practices rather than idealized notions of data discovery; and examining the relation of data discovery to a web of other practices and sociotechnical elements.

Koesten et al. (2017) discuss the information-seeking process for data on the web including influencing factors in a framework for interacting with structured data (Fig. 2.2). This is also based on information-seeking models and presents another perspective on the conceptualisation of data-centric tasks, search strategies, as well as evaluation and exploration activities in data search. The pillars of this framework are: *tasks, search, evaluation, exploration* which together enable or can lead to *using data*.

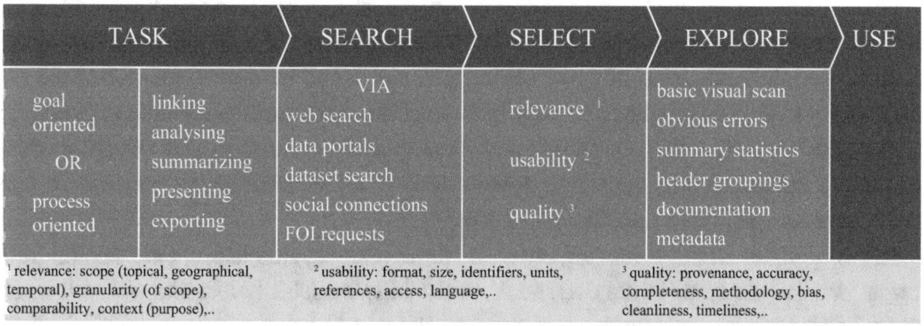

TASK		SEARCH	SELECT	EXPLORE	USE
goal oriented OR process oriented	linking analysing summarizing presenting exporting	VIA web search data portals dataset search social connections FOI requests	relevance [1] usability [2] quality [3]	basic visual scan obvious errors summary statistics header groupings documentation metadata	

[1] relevance: scope (topical, geographical, temporal), granularity (of scope), comparability, context (purpose),..

[2] usability: format, size, identifiers, units, references, access, language,..

[3] quality: provenance, accuracy, completeness, methodology, bias, cleanliness, timeliness,..

Fig. 2.2 Framework for human structured-data interaction. The authors recognize this is not a linear process but often involves circular activities between the pillars. Adapted from Koesten et al. (2017)

The framework, which is further discussed in Chap. 3, is based on a series of qualitative studies and aims at helping system designers and publishers of data to understand what people do when searching for and engaging with data. They differentiate between different selection criteria, focusing on (i) *relevance*, including the content of the data, their topic as well as temporal and geographical scope; (ii) *usability*, considering practical implications of the data that influence how they can be used, such as format, size, documentation, language, comparability (e.g., identifiers, units of measurement) and access (e.g., license); and (iii) *quality*, referring to dimensions that allow users to judge the data's condition for a task, such as their completeness or accuracy.

The concept of *fitness for use* is sometimes used in a similar context to refer to quality, emphasizing customer needs and how these are met by the specifications of a product (Bishop et al. 2018). The term can be traced back to 1950 in a quality management book still available today (Juran 2017). The key notion is the task dependency of quality assessments, something which has also been the finding of the authors' recent work on data-centric sensemaking (e.g., Koesten et al. 2021).

Ramdeen's doctoral work, also discussed in Chap. 3, applies Ellis's model of information-seeking to look at the information-seeking behavior of geologists when searching for physical samples (such as fossils, and rocks) which are considered primary sources of data (Ramdeen 2017). There, data are also described as having different characteristics and access points compared to formally published resources. Information resources used to discover data include personal contacts, reviewing data citations in publications or other literature, and browsing databases.

Friedrich also makes the argument that data-related information behavior is not covered by drawing just on insights from literature-related information behavior studies (Friedrich 2020). Her work looks specifically at social science survey data, drawing on expert interviews and surveys. Her findings emphasize the role of communities in facilitating data discovery, the role of expertise, and the importance of documentation.

A key commonality across studies which are rooted in information-seeking work is that **data discovery affords certain data-seeking strategies and user needs that are unique to data**. Certainly not ALL of these are unique, and it can be argued that when the data discovery process is described at a very high level the steps from realizing an information need to taking action in "searching" and subsequently evaluating and selecting search results are similar, if not the same. Nonetheless when looking at this process in more detail, aspects specific to data begin to surface.

As we will see in the following chapters, these data-specific aspects are related to the format and the structure of the data, such as machine-readability and size, as well as elements related to the environment the data are situated in, such as access restrictions and licenses. There are also aspects on the users' side, such as data-specific tasks, required skills for working with data, or issues of understandability both within and between disciplinary domains. Other aspects concern the data discovery "system", such as interfaces and the metadata that facilitate some of these processes.

Considering these aspects together, it becomes obvious that we need data-specific discovery tools rooted in studies of data-specific discovery tasks, needs, and strategies. Enhancing the current state of the art entails expanding search paradigms tailored to textual information to include data-specific behaviors.

2.2.2   Data Discovery Versus Data Seeking Versus Data Search

In this work, we describe *data discovery* as the entire process of establishing a data need, searching or otherwise encountering data, and potentially selecting a dataset, including refinements of the search and evaluating the data's fitness for purpose. It is important to note that when we discuss data discovery we do not think about looking up single values or answering questions based on data (as in Hirschman et al. 2001), but rather focus on discovering a collection of data about something.

We take a dynamic view of the data-seeking process, including serendipitous encounters with data, and do not limit *data-seeking* to a goal directed activity only. We differentiate the term *data search* (or *data retrieval*, used synonymously here), as the actual search activity which takes place during a data discovery process where individuals actively engage with search systems to locate data.

This differentiation draws on Wilson's nested model of information-seeking behaviors and information search behaviors (Wilson 1999). In his model, research into information search behaviors, seen as a subset of less goal-oriented information-seeking, focuses on interactions between users and search technologies. As mentioned, data discovery is rooted in information-seeking, which generally refers to activities undertaken to acquire information to fulfill a need, want or gap in knowledge (Case and Given 2016).

2.3   What Do We Mean by Human-Centered?

Human-centered is a term used in different ways; here, we refer to thinking about data discovery from the perspective of the person(s) engaging in the discovery activity. The focus of interest is on the interaction process and the "user" experience taking into account different contexts and data needs.

We discussed user-centered models in information science, where a key point is that information discovery is defined through interactions, both between users and systems, but also interactions between contexts and users. A shift in information science from a more system-oriented view towards the standpoint of the user, or the information seeker (Dervin and Nilan 1986), started roughly in the mid-1980s (Tang et al. 2019; Savolainen 2007). This is also reflected in the prominence of user-centered design in computer science and software development (Kashfi et al. 2017), where the term "user-centered" or "human-centered" commonly describes an approach to the development of interactive systems focusing on making them usable by drawing on multiple disciplines (Heimgärtner 2020).

Despite its prominence, the explicit focus of the "user" in technology development has its drawbacks. We acknowledge the limitations of user-centered models drawing on Van House (2004), including defining an individual purely in terms of their interactions with an information system but not within their different contexts and with other individuals in shared or distributed actions (Talja and Hansen 2006). Further, persons who do not interact with discovery systems, "non-users", are not typically represented, limiting the generalisability of such models. Thinking about non-users and their data needs can help to contextualize the human perspective and also to inform the design of discovery systems (Wyatt 2003).

However, especially for data discovery, we believe it is important to emphasize the "human" not by focusing on specific people, or even types of expertise, but on data needs and contexts surfacing when examining data discovery in-situ, as we discuss further in the following chapter.

2.4   Data Discovery in Practice

We close this chapter by providing examples drawn from our own research about the diversity of data needs and data discovery practices which we have observed in our empirical work.

We spoke with various researchers during the course of interview studies, including Nayla,[5] a water resources researcher who was working to create a model describing water flow for an area in southeastern Asia. She relied completely on data created by other people to do her work. Her **data needs** included digital surface elevation maps, local and historical data about rainfall, population, and streamflow, and other satellite data and

---

[5] Participants' names and some details have been altered to preserve anonymity.

code that she could adapt to her model. Her data needs were also very specific. They were restricted to a particular geographic location, and needed to meet the specific requirements for her model.

Not all of the data that she needed for her work were findable. Some of these data did not actually exist, because they had never been collected or were not collected at the needed level of geographic granularity. Other data were stored physically in a records office at the government or were available online, but were described and made searchable using different standards and interfaces than what she was accustomed to.

Jan, a paleontologist, also experienced similar difficulties. He needed physical fossils of shrimp claws as he worked to understand the fossils' material composition and trace the development of the species over time. Jan learned of possible specimens for analysis through email discussion lists and via books with historical illustrations. He followed clues from these sources to locate specimens held in museums. He then consulted with museum curators to locate the fossils, as they were possibly hidden or unlabelled, becoming findable only through the expertise of the museum curator.

Celeste, another interview participant, sought data to use in new ecological experiments, for teaching and also in simulation models. She **interacted with different data sources** to find the data she needed for these purposes. She looked in the academic literature, browsed through data repositories, performed quick searches in Google, and, most importantly for her, talked to her direct network of peers. She also followed up with data creators, not just to gain access to data, but for social reasons such as laying the groundwork for eventual collaborations.

The story of Peter, a soil scientist working on a collaborative research project, exemplifies the **data-centric sensemaking practices** which we present in Chap. 5. Peter located data created by other people working on his project team in a shared spreadsheet, where the data were organized according to who collected the soil samples at different catchment areas. When scrolling through the data and checking columns against each other, he discovered an unexpected value and needed to track down lab members responsible for data collection in order to understand possible sources of error in this shared dataset. He also needed to consult the data documentation about how data were collected, and under which conditions, information which in this case was embedded with the spreadsheet itself.

Researchers are not the only people who engage in data discovery and sensemaking. We can also imagine the scenario of a journalist, Alex, writing an article on how Hurricane Sandy impacted the gasoline prices in New York City in the week after the incident. For Alex, two datasets, one from the American Automobile Association (AAA) and one from Twitter could both potentially be relevant. Both datasets document the gasoline available for purchase in New York City in the correct time period. The choice of which dataset to use depends on the specifics of his data need, potentially the purpose and requirements of algorithms or processing methods, and his level of data literacy. If he wanted data which were ready-to-use, with fewer time restrictions, the curated AAA dataset may be the best fit. If, however, his purpose were to establish a very detailed timeline, the messier Twitter data could be a better match, provided he has experience with the necessary data cleaning tools.

There are many reasons to search for data, also amongst interested citizens. Imagine Anna, a young Austrian, who was writing an essay for school about voting behavior in general elections over the past years. She wanted to compare demographic and socioeconomic characteristics between voters and non-voters and analyze these insights by voting district. In planning her analysis, she decided that it would be interesting to broaden the scope of her data to include several countries to identify commonalities and differences. Her questions quickly required multidimensional data, which might not be available in existing textual articles. Without having tried, she feared that this would require several data search activities, per country and potentially involve merging differently structured datasets.

These practical examples serve to contextualize the remainder of the chapters in this lecture, beginning with a discussion of individuals' *data needs* in Chap. 3.

Data Needs

3

This chapter synthesizes findings about the data needs of individuals in different communities. After defining the concept of data needs (Sect. 3.1), we discuss users of data and data discovery systems (Sect. 3.2). We then review common methodologies employed to study data needs (Sect. 3.3.1) and examine existing work through three lenses identified from the literature (Sect. 3.3.2). We end the chapter by reaching four conclusions about data needs (Sect. 3.4), proposing that thinking about data needs "in situ", as local happenings influenced by an individual's purposes and uses for data, can be fruitful.

3.1 From *Information Needs* to *Data Needs*

The concept of *information needs* forms the backbone of much existing work investigating information-seeking behaviors. Such work, including the models discussed in Chap. 2, posits individuals who engage in information-seeking activities in order to satisfy a particular information need.

The study of information needs stretches back to the 1960s (see Taylor 1968). While an extensive review of the literature on information needs is beyond the scope of this chapter, two primary perspectives can be observed. Users and their information needs tend either to be conceptualized as passive recipients of objective information or through a more active, constructivist perspective (Dervin and Nilan 1986). Particularly in the constructivist view, information needs, which can be both explicit and implicit, stem from an individual's efforts to make sense of the world (Dervin 1983; Wilson 1994) and evolve in response to a number of factors including local and changing contexts (Leivrouw 2001), information discovery activities (Bates 1989), and an individual's tasks and purposes (Ingwersen 1992).

© The Author(s), under exclusive license to Springer Nature Switzerland AG 2022
K. Gregory and L. Koesten, *Human-Centered Data Discovery*, Synthesis Lectures on Information Concepts, Retrieval, and Services,
https://doi.org/10.1007/978-3-031-18223-5_3

We extend the concept of *information needs* to the search and discovery of data, and define a *data need* as **data which individuals desire or require, for a particular purpose**. In line with constructivist views of information needs, data needs may also be explicit or implicit, are locally defined, and evolve over time.

Individuals seeking data also need information *from* and *about* data, which they use to evaluate the data they find and to place data in context. This type of *contextual information* about data can be provided formally and informally through different forms of data documentation, as well as through social channels. As such, the contextual information surrounding data is a vital component of human-centered data discovery, which we address in detail in Chap. 5 in our discussion of evaluation criteria and sensemaking practices.

3.2   Who Needs and Uses Data?

In 2014, more than six million data professionals, people whose primary job activities include collecting, storing, managing and analyzing data, were spread across all industries in the European Union (Cattaneo et al. 2015). This number continues to increase today, and includes people who need data for various purposes across a variety of sectors and backgrounds.

Researchers in domains ranging from the arts to medicine to zooarchaeology need data created by other people, throughout phases of their work (Borgman 2015; Faniel et al. 2018). Journalists need data to craft and support their messages (ProPublica[1] or The Bureau of Investigative Journalism[2]); data scientists need data to create models and make predictions[3]; and public audiences need data to understand, among other things, medical claims published in newspapers. Individuals seeking data can inhabit multiple such roles simultaneously. A graduate student, e.g., may also identify as a researcher, a data scientist and a member of the general public.

Rather than belonging to a single "user group" or "profile", users of both data and of data discovery systems can be thought of as belonging to multiple *data communities* simultaneously. These communities are not solely defined by an individual's disciplinary expertise. They can form around shared data pools or common methodologies, i.e. data science, modeling, or methods used in the digital humanities (Gregory 2021; Leonelli and Ankeny 2015). Each of these data communities may have different norms about how data are reused, described, and made available. This in turn can shape an individual's data needs and their data discovery practices. This also exemplifies the dynamic nature of expertise in relation to data tasks (Koesten et al. 2021).

---

[1] https://www.propublica.org/.

[2] https://www.thebureauinvestigates.com/.

[3] https://huggingface.co/.

The use of technological systems exists on a spectrum, rather than being a binary activity (Wyatt 2003; Oudshoorn and Pinch 2003). At times an individual may be a user of data or of a data search system, while at other times they may not. As with reasons for *use*, reasons for non-use vary. Individuals may not have access to data or necessary technologies; they may not have the correct set of skills, or data reuse and data seeking may simply not be relevant to their particular context (Gregory 2021). Understanding use and its relation to non-use can help to nuance our current understandings of users, data needs and data-seeking practices (Wyatt 2003).

In the same vein, certain groups are more represented than others in the data discovery and reuse literature. This also biases our understanding of users and their data needs. Much existing work has focused on the data needs of researchers in a core set of fields such as biomedicine, the environmental sciences and the social sciences (Gregory et al. 2019). The data practices of arts and humanities researchers, on the other hand, are much less frequently addressed (Edmond et al. 2017).

Other potential users of data, for example the general public or citizen scientists, are also not well represented in the literature. What little information exists about the data needs of public audiences is primarily found in the context of other studies focusing on the use of particular data portals. For example, in a study of the national Dutch data infrastructure DANS,[4] Borgman and colleagues spoke with a data consumer who searched for archaeological data in order to prepare for his hobby of giving historical tours (2019). Similarly, a user study conducted at the Australian Research Data Commons surfaced individuals searching for genealogical data while researching family histories (Wu et al. 2021; Liu et al. 2021).

Although there is a growing body of research on citizen scientists (Silvertown 2009; Bautista-Puig et al. 2019), not much attention is paid to how those individuals are discovering data and using them for their own purposes. Citizen scientists are normally portrayed as contributors who aid in data collection and annotation efforts, i.e. in the Zooniverse platform where volunteers classify or label data according to guides or tutorials (see Simpson et al. 2014). While there are other literatures to draw from, including work in data activism (Kennedy 2018) which provide evidence of citizens both acting as data users and contributors, the data needs of these individuals have not yet been systematically studied.

3.3   Which Data Are Needed for Which Purposes?

3.3.1   Methodologies for Researching Data Needs

Commonly used methods to investigate data needs include extracting data needs from textual information and methods with a closer engagement with data seekers, such as

---

[4] https://dans.knaw.nl/.

interviews and ethnographic approaches. In this section, we do not aim to provide a comprehensive review of all studies using these methodologies, but rather we give examples of such studies and their findings.

3.3.1.1 Method 1: Extracting Data Needs from Textual Descriptions

Approaches such as search log analyses or the use of question corpora are one way of abstracting information about data needs. Kacprzak and colleagues, e.g., conducted a study combining a detailed search log analysis and an analysis of submitted textual data requests to four open government data portals in the UK, Canada and Australia (2019). The study found that queries for data tend to be shorter than general web search queries, and that they contained more words related to geospatial and temporal information, as well as data format and file type. The authors also suggest that searching for governmental data is a task that tends to be work-related, as most of the searches occurred during working hours in the respective countries and from desktop computers.

Borgman and colleagues conducted a log analysis at a data repository for the social sciences and humanities in the Netherlands, analyzing transaction logs combined with a database of registered users over a three-year period (2015). While the analysis provided a high-level descriptive picture of the use of the data archive, the logs were difficult for the authors to interpret, as no information was provided about which part or parts of a downloaded dataset were actually needed. The authors emphasize that log analyses on their own are limited, as they do not provide information about the motivations and purposes underlying data search.

Collections of expert-sourced question corpora have also been used as a means of extracting and classifying data needs within a particular domain of research. Such corpora are created either through the collection of actual user requests, as in genomics (Hersh and Voorhees 2009), or through more simulated tasks where experts are asked to generate "real life" questions following specified templates and guidelines, as in the BioASQ challenge (Nentidis et al. 2017) and the bioCADDIE Dataset Retrieval challenge (Roberts et al. 2017). Once the corpora are collected, entities or nouns are extracted from the questions. These entities are then classified and annotated to create a typology of data needs which can be used to inform the development of data search tools and services.

Exemplifying this approach, Löffler and colleagues created a question corpora based on questions elicited from 73 biodiversity scholars across Germany (2021). Using this corpora, they identified categories important to biodiversity researchers searching for data, which included information about species, biological and chemical processes, materials and chemicals, and data parameters. They further analyzed existing metadata schemas used at data repositories to identify a gap between the specificity of these data needs and the metadata schemas which commonly are used to support data search.

Kato and colleagues also incorporated a question corpora in their DataSearch track at the NTCIR-15 conference, in which six research groups tested their solutions for an ad-hoc retrieval task for data search (2020). Rather than focusing on a particular disciplinary domain, the focus of the study was on government statistical data. The topics used in the task were drawn from the database of a community question-answering service which included links to statistical data portals in Japan. In this approach to identifying data needs, workers from crowd-sourcing pools were asked to translate topics such as "How many people have a second house" into queries and to make relevance judgements about the data found using such queries. While crowdsourcing such work is a common approach in developing information retrieval tools (Maddalena et al. 2017), it arguably does not reflect the practices and evaluative decisions of actual data seekers.

3.3.1.2   Method 2: Surveys

Surveys have been used as a method to engage with data seekers and identify their data needs at a deeper level. Such surveys have been embedded within data repositories and portals to capture data seekers as they are searching for data, but they have also been conducted as stand-alone studies.

The Global Biodiversity Information Framework undertook a survey designed to understand more about the data needs of individuals accessing the data portal (Ariño et al. 2013). A 21-item multi-lingual questionnaire was distributed via disciplinary mailing lists, networks and professional associations. Drawing on more than 700 individual responses, the authors found that respondents need data of various types (i.e. taxonomic names, species information, population density) for various uses, i.e. for performing basic science, which was selected most often, applied science (i.e. conservation planning and management), and to address societal issues (i.e. ecotourism or recreation).

The Research Data Australia discovery service takes a different approach to surveying individuals using their data portal. In an ongoing study in 2022, individuals visiting the portal are shown a 2-question pop-up survey in which they are asked to describe their professional role as well as the data which they need. Respondents to this survey are then asked to complete a longer questionnaire designed to elicit further context about their data needs and search practices. Initial results for the short survey indicate a strong positive correlation between individuals identifying as researchers and the specificity of their data needs. Weaker positive correlations have also thus far been identified between students/industry employees and more general data needs (Wu et al. 2021; Liu et al. 2021).

In another survey study, Gregory and colleagues aimed to capture information about data needs and data discovery practices in academia, across disciplines and data sources (Gregory et al. 2020a). They distributed a questionnaire to authors who had published an article indexed in the Scopus literature database in the previous three years. The questionnaire asked respondents to describe their data needs in an open-ended question as well as to classify the type of data which they seek. While the majority of the respondents selected needing observational data in the course of their work, approximately fifty percent of respondents also selected needing more than one type of data. Slightly more than half of respondents across domains reported needing data outside of their disciplinary area of expertise.

3.3.1.3 Method 3: Interviews and Ethnographies

Semi-structured interviews, observational studies and ethnographic fieldwork provide rich detail about data needs and data practices in different communities.

Krämer and colleagues, e.g. conducted semi-structured interviews following an observational study combined with a think-aloud task to investigate the data-seeking behaviors of social scientists (2021). The interviews were used to add nuance to the observational studies in which participants sought data on topics of their own choosing. The topic of these searches were triggered in response to participants' past research interests, to verify or disprove a hypothesis, or to identify unexplored variables of interest in a popular dataset. The granularity of the participants' data needs evolved over time and in response to their observed searching behaviors (Krämer et al. 2021).

Gregory and colleagues also conducted semi-structured interviews with individuals who had visited Elsevier's DataSearch platform.[5] They found a great variation in data needs among participants, including needing data about superconducting temperatures, spectral data, field observations and social media posts (Gregory et al. 2020b). Data needs were found to change as research interests, job changes and new collaborations were formed.

Interviews which uncover details about data needs are not always conducted within studies about data discovery per se. The extensive ethnographic work of Borgman and colleagues provides one example (2021). For example, Borgman and her colleagues spent months with the Center for Embedded Networking and Services, an interdisciplinary team of ecologists and engineers, conducting interviews, performing observations, examining documentation and participating in meetings (ibid.). Their focus was to study practices of data collection, use, storage and sharing; they also identified information about the data needs of researchers. Ecologists working on the project, e.g. needed data about weather, solar, and river observations, as well as remote sensing and demographic data in order to verify and contextualize data which they create and to calibrate their own instruments (Wallis et al. 2013; Wynholds et al. 2012).

3.3.2 Common Lenses for Analyzing Data Needs

There are common lenses or frames which are used to discuss data needs in the literature. Such lenses include examining the needs of individuals belonging to particular disciplinary groups, of those working with particular types of data, or of those using data for a particular purpose.

---

[5] DataSearch now exists within other Elsevier products, Mendeley Data and Data Monitor, rather than as a separate platform.

In this section, we summarize and give examples of the findings resulting from analyses conducted using these framings, highlighting where these lenses connect and overlap. We then close the chapter by drawing on our analysis of both the methodologies used to study data needs (Sect. 3.3.1) and these common lenses to reach four overarching conclusions.

3.3.2.1   Lens 1: Data Needs According to Disciplinary Domains in Academic Research

Disciplinary domains provide the most common frame for analyzing data needs in studies of scientific or research data. This reflects the extensive literature documenting the influence of disciplinary norms on data practices (i.e. Leonelli 2016) as well as on scientific knowledge production and scholarly communication (Kuhn 1962; Knorr-Cetina 1999).

Dixit and colleagues conducted a two-phase user-study combining interviews followed by usability testing to determine the data needs of *biomedical researchers* as part of the development of DataMed, a data discovery index for biomedical data (2018). Biomedical researchers in this study included individuals in various sub-disciplines, i.e. molecular biology, genomics, and public health, who all had experience analyzing biomedical data. The study concluded that biomedical researchers need data from different sources, which have metadata describing biomedical concepts, data types (i.e. gene expression data or clinical data), and details about data collection and processing, a finding in line with other studies which we further discuss in Chap. 5. The data needs of these biomedical researchers were found to be multidimensional and shaped by an individual's technical savvy and experience.

In another study utilizing a disciplinary framing, Ramdeen builds on the information-seeking models proposed by Ellis and colleagues (Ellis 1989; Ellis et al. 1993; Ellis and Haugan 1997), to examine the information-seeking practices of *geologists* looking for physical samples, i.e. cores, cuttings, fossils, and rocks, which serve as a primary data source in the geosciences (2017). These physical forms of data, along with their attached metadata and documentation, can be maintained and reused when given appropriate care and resources. When these samples and their metadata are not maintained, they may not be discoverable or found by researchers who need them (Ramdeen 2017).

In a recent study of *social scientists*, another oft-studied disciplinary group, Papenmeier and colleagues aim to isolate the "genuine information needs" of social science researchers during their initial stages of seeking data (2021). Using a questionnaire, the authors asked respondents to express their data needs as they would to a colleague in an informal situation. These descriptions were then segmented and placed into three categories describing different parts of respondents' data needs: topic (i.e. politics, economy, psychology/personality); "meta" information (i.e. sociodemographic details or geospatial attributes); and intention, or what a researcher wants to discover about a particular topic (i.e. attitude, perception or satisfaction). The authors conclude that the specificity of the data needs expressed in these descriptions do not match the more generic queries used to search for data.

Analyzing data needs solely through a disciplinary lens does not take into account interdisciplinarity; in their survey investigating data discovery practices, Gregory and colleagues found that over fifty percent of researchers identify with more than one disciplinary domain. Interdisciplinary projects also create new communities and contexts which blur disciplinary lines and necessitate negotiating new norms about data reuse (Gregory et al. 2020a).

3.3.2.2 Lens 2: Data Needs According to Data Type

Data needs are also framed in terms of the *type* of data which individuals need. Data types are defined in different ways, ranging from (i) the methodology used to collect data to (ii) the data's structure and format.

Data Type Determined by Data Collection Methodology

Existing work (Gregory et al. 2019, 2020b) draws on broad categories of data collection methodologies proposed by the National Science Board (National Science Board 2005) to classify needed data as *observational, experimental, or computational.*

According to this classification, *observational data* result from recognizing, recording, or noting occurrences. These data are often produced with the help of instruments, and include data such as weather observations, polling data, photographs, maps, and economic indicators. This type of data is usually tied to a specific time and place, and cannot be easily replicated. *Experimental data* result from procedures carried out in controlled environments that are designed to test hypotheses. Examples include chemical reaction rates, genomic expression data, and data collected via collider experiments, electron microscopy, and controlled psychology experiments. *Computational data* are produced by running computer models or simulations. Computational techniques are used in the physical, life, and environmental sciences, but can also be used in social sciences such as economics (Borgman 2015; National Science Board 2005).

An analytical literature review analyzed the data needs of users of observational data to identify commonalities across disciplinary contexts (Table 3.1; Gregory et al. 2019). Descriptions about data needs in the reviewed literature include information about the instruments and technologies used to collect and record data and the spatial and temporal resolutions and formats used to capture observations. The authors also linked these descriptions of data needs to the documented purposes for using data, including purposes used to drive new research, so-called foreground purposes (Wynholds et al. 2012; Wallis et al. 2013), and purposes supporting academic work, or background purposes.

This same type of analysis could be done for other data types according to this classification, i.e. for experimental data or computational data. Descriptions of the experimental data needs documented in the literature center on genetic data, where researchers need information about genetic markers, sequencing data and protein structures (Brown 2003; Key Perspectives 2010). Genetic data is often recorded in microarrays which measure the expression of thousands of genes at the same time (Zhang et al. 2010); at the same

Table 3.1 Descriptions of the documented data needs of users of observational data

Users in this community…	Need this type of data	For these purposes (italicized = foreground, normal = background)
Astronomy	Data from sky surveys, telescopes, archives, repositories, data catalogs, virtual observatory systems	*New questions of old data,* baselines, instrument calibration, physical properties, model inputs, data integration
Earth and environmental sciences	Plant, animal, water, weather, solar observations; soil analyses, rock thin-section and satellite images; maps, geographic, demographic and census data; continuously collected and transmitted data, data at temporal/spatial scales, raw and summarized data	*New questions of old data, meta-analyses,* calibration, context, baselines, reference, model inputs, verification, comparison, environmental planning, policy- and decision making, education, instrument monitoring; data integration
Biomedicine	Images, complete fMRI studies, pathology results, patient observations and demographics; population-level disease data, behavioral data	*Disease/disorder research, new visualizations,* evaluations, 3-D anatomical pictures, preparing research outputs, education, patient care
Field archaeology	Field notebooks, photographs, artefacts, stratigraphic baselines; data at temporal/spatial scales	*New insights from data aggregation,* comparison, triangulation; training, dissertations, assignments, preparing tours, inventories of local excavations
Social sciences	Survey data (often only one question is of interest), long-running datasets/surveys, interviews, archival documents, images, videos	*Re-interpret datasets; new questions, comparative research,* comparison, preparations, training, dissertations

Adapted from Gregory et al. (2019)

time, researchers using genetic data also need a broad view of the entire genome (Megler et al. 2015). Despite this example, there are many more descriptions of observational data needs in the existing literature as compared to other data types.

Specific collection methodologies, i.e. data collected through surveys, are also used to frame discussions of data needs. Friedrich examines the data needs of social science researchers seeking survey data in terms of individuals' experience levels and goals (2020). She finds that individuals with less experience with survey data tend to have more basic goals for using data (i.e. to practice analysis techniques) as compared to experts in survey research who have more ambitious goals, i.e. to publish new studies in academic journals.

She also emphasizes that survey data reusers need extensive documentation, including codebooks and technical information, due to the structure of the survey data themselves, which consist of matrices with variable values in columns and cases arranged in rows (ibid.). The method of collecting and recording survey data influences the ultimate shape and structure of the data.

Data Type Determined by Data Structure

The structure and format of data, particularly structured data, is another common frame used to discuss the type of data which individuals need. Structured data, often associated with quantitative data, are organized in highly regular ways, i.e. columns and fixed fields, and are recorded in databases, spreadsheets and web tables (Losee 2006). Unstructured or less structured data are often linked with qualitative forms of data and data collection and do not have the same type of explicit "structuring" information. Such data can include text or also audio and video recordings.

Studies examining the need for structured data are much more common than those examining less structured, qualitative, data needs, perhaps because of longstanding debates about the ethical considerations of reusing qualitative data (see Bishop 2009). Building on an analysis of data usage at data archives in Finland and the United Kingdom, Bishop and Kuula-Lummi suggest that this may be changing and that the reuse of qualitative data is becoming more common (2017). The authors of this study identify three main purposes for the reuse of qualitative data in the analyzed archives: research, teaching, and learning. They suggest that data collections used for teaching tend to include multiple types of data (i.e. transcripts, oral histories, etc.) which are of potential interest to young people. Data used for research in their study were related to broader topics (i.e. health, food, and work). The topic of data was not always a driving factor in reuse, as when individuals needed data of a particular type to inform the development of a research plan.

Koesten and colleagues conducted interviews with individuals who work with *structured data* in their daily jobs, including scientists, data analysts, financial traders, IT developers and digital artists (2017). Interview participants reported needing both openly available data and proprietary data in a variety of thematic areas such as environmental research, criminal intelligence, retail, and transport. Similarly, the authors found that participants used data for a range of purposes and tasks, which we discuss further in the next section.

Many studies about open government data focus on the use of structured data. Xiao et al. identify major challenges for users of open government data, which include discovering, accessing and understanding data. These challenges are due in large part due to a lack of contextual information and adequate metadata (Xiao et al. 2020a). In another study focusing on open government data, Koesten et al., propose a template of nine questions to ask of a dataset in order to create a summary for the purpose of selecting data for a particular need (2020).

Degbelo synthesizes 27 user need statements drawn from the literature as a step toward creating a taxonomy of user needs for open data (2020). The author clusters the needs extracted from these statements into ten categories based on how individuals use (or intend to use) the data and relevant contextual information. These categories include assessing relevance, establishing trustworthiness and assessing usability. While this taxonomy is mostly related to the contextual information surrounding data, it reflects a broader trend to conceptualize data needs in terms of the *purposes* for which data are used, rather than in terms of specific user profiles.

3.3.2.3 Lens 3: Data Needs According to Purposes, Uses and Tasks

Other recent work proposes that focusing on the purposes, tasks or uses to which data are put can provide a community-spanning way to conceptualize data needs (Koesten et al. 2017; Xiao et al. 2020b). This work builds from the premise that data-centric search activities likely depend on the tasks for which people need them (see Ingwersen 1992); this perspective is also apparent in the earlier descriptions of data needs in this chapter, which are closely linked with purposes or tasks for which data are used.

While there are a variety of task classifications related to information-seeking (i.e. Vakkari 2005; Toms 2013), there is a relative dearth of corresponding taxonomies for data-centric work tasks, perhaps reflecting changing work practices over the past decade (Koesten 2019). Existing task classifications include tasks related to data science activities (Pfister and Blitzstein 2015); data analysis tasks performed by individuals in business (Convertino and Echenique 2017); and the sub-tasks "data workers" engage in when encountering uncertainty during data analysis (Boukhelifa et al. 2017). These classifications do not include a deeper discussion of data-centric work tasks; in an attempt to fill this gap, various frameworks for understanding the tasks and purposes influencing data needs and data seeking have been proposed.

Building on the results of a mixed-methods study with data professionals, Koesten et al. propose a typology of work tasks for data, which includes process-oriented and goal-oriented tasks (2017). In process-oriented tasks, data are used for something "transformative," i.e. in machine learning processes, integrating data into a database, or when visualizing data. Goal-oriented tasks for data include purposes such as seeking the answer to a question, finding patterns, or making comparisons between datasets and data points (Fig. 2.2 in Chap. 2). One of the primary differences between these two categories lies in the contextual information about data which users need in order to perform these tasks. For process-oriented tasks, contextual information about timeliness, licenses, data quality and methods of data collection have a high priority; for goal-oriented tasks, attributes such as coverage and granularity are important.

The authors further differentiate between five types of data-centric tasks within these broader categories: linking; analyzing; summarizing; presenting; and exporting. *Linking* tasks involve finding commonalities and differences between different datasets, requiring the ability to view multiple datasets at the same time. *Analysis* tasks center on identifying trends over time or making predictions, while *summarizing* involves creating different representations of data and *presenting* includes creating visualizations or textual descriptions of data. Finally, *exporting* refers to tasks involved in sharing or publishing data, i.e. creating metadata or posting data in repositories. Data-centric tasks are complex; individuals often engage in all five types of these activities as they perform goal- or process-oriented tasks (Koesten et al. 2017).

Xiao and colleagues apply Koesten et al.'s framework for data tasks to identify common tasks across disciplines for individuals working with structured research data (Xiao et al. 2020b). Based on the findings from semi-structured interviews with researchers, the authors identify six tasks, classified as either data-driven tasks, in which research questions emerge from engaging with data, or model-driven tasks, in which data are specifically needed to answer a particular research question defined in advance.

Wynholds and colleagues also propose a way to conceptualize different uses for research data, although at a broader level (2012). *Background uses* for data, i.e. making comparisons, benchmarking, and instrument calibration, are undertaken to support research; *foreground uses*, which drive new research, are often limited to reports of "asking new questions of data" (Wynholds et al. 2012; Wallis et al. 2013), and are not as well-documented (Gregory et al. 2019). In their survey investigating practices of data discovery, however, Gregory et al. found that the majority of respondents (71%) selected using data as the basis for a new study, a foreground use (Gregory et al. 2020a). Other common purposes for using data among survey respondents included experimenting with new methodologies and techniques such as developing data science skills or completing particular data-related tasks, i.e. trend identification or creating data summaries (as suggested by Koesten et al. 2017).

Using the results of this survey and existing research life cycle models, the authors propose a typology for the uses of data, spanning phases of academic work (Gregory et al. 2020a). This typology includes five classifications for using data: *data reuse* (using data as the basis for a new study); *project creation and preparation*, i.e. preparing a grant proposal; *conducting research*, which includes tasks such as using data to calibrate instruments or as model inputs; *data analysis and sensemaking*, including tasks such as data visualization and trend identification; and *teaching* (ibid.). The authors stress that these uses are interwoven with each other and other types of research practices and recognize that the typology does not account for other roles which data play, i.e. stimulating collaborations and creativity.

Pasquetto and colleagues propose another typology for research data reuse that places types of data use on a continuum between comparative uses, i.e. ground truthing or instrument calibration, and integrative uses, i.e. bringing together data for new analysis or to identify new patterns (2019). They find that integrative uses occur very rarely, perhaps once in a research career, whereas comparative uses occur fairly frequently. Integrative uses, which they classify as emergent practices, require more expertise, often necessitating collaborations with data creators, as opposed to more routine comparative data uses.

3.4 Conclusions About Data Needs

In this chapter, we have identified common methodologies and lenses used to study and frame discussions of data needs, providing examples of the data needs of individuals in different communities and contexts along the way. Based on this synthesis, we arrive at the following conclusions about data needs and how they have been studied thus far.

Data Needs Are Specific, Multiple and Diverse

Across the reviewed literature, we see that data needs are specific, multiple and diverse. Observational data users, for example, need data from particular locations (be those geographic, anatomical or astronomical) at particular resolutions or collected with particular instruments (Gregory et al. 2019). Researchers across disciplinary domains have highly specific data needs (Wu et al. 2021), and data professionals, including data scientists, data journalists and data analysts, use data for specific process- and goal-oriented tasks (Koesten et al. 2017).

Data seekers also require multiple types of data for multiple purposes, which reflects the complexity and iterative nature of working with data (Pasquetto 2019; Koesten et al. 2021). The specificity and multiplicity of data needs creates a diverse spectrum of needed data, reflecting the view that what serves as data for one individual within a particular epistemic community will not count as data for another (Leonelli 2015; Borgman 2015).

Data Needs—and Classifications of Those Needs—Are Dynamic and Overlapping

Although we have presented data needs through common, separate lenses in Sect. 3.3.2, our analysis also shows that it is not possible to fully understand the complexity of data needs using just one lens alone. Data needs are shaped by relationships between disciplinary domains, data types and structures, and the uses to which data are put.

Data needs are dynamic and evolve in response to searching behaviors (Krämer et al. 2021), but also in response to broader changes in data practices within communities. Similarly, certain types of data uses and tasks, particularly integrating data from multiple sources, may be emergent practices which require a high level of collaborative effort and thus do not occur that often (Pasquetto et al. 2019). Our synthesis also shows that classifications for data needs and data tasks or uses are themselves continuing to develop and change.

Data Needs are Intertwined with Contextual Information Describing Data

As stated in the beginning of this chapter, the contextual information surrounding data is a critical part of human-centered data discovery, particularly in terms of evaluating data for reuse and understanding. We also see that contextual information about data is intertwined with and shapes individuals' data needs. A lack of available metadata and contextual information hinders how data seekers discover, access, evaluate and understand data (Xiao et al. 2020; Friedrich 2020); individuals therefore may not use (or be able to find) data without such contextual context.

Data Needs Are Determined and Best Conceptualized In Situ

Data needs are informed by local, dynamic contexts. These contexts involve combinations of relationships between, e.g., an individual's role, disciplinary and professional norms, and the intended purposes, tasks or uses for data. Data needs are also shaped by the availability of data, levels of expertise, and the point in time at which data are used within workflows. We argue that data needs are therefore best construed in situ, as local configurations or happenings, and that this conceptualization should be the starting point for the design of data discovery systems.

Discovering Data

4

This chapter reviews four common sources used to discover data, from both a technical and interaction perspective: search engines (Sect. 4.2), data portals and repositories (Sect. 4.3), academic literature (Sect. 4.4), and social connections (Sect. 4.5). The chapter ends (Sect. 4.6) by drawing four conclusions about how people discover data in these sources.

4.1   Common Data Sources

Finding data to meet an individual's data needs is often challenging (Xiao et al., 2019), in part because data are distributed across different sources, including general and specific data repositories, data markets, open (governmental) data portals, project websites, museum archives, and general-purpose resources, i.e. Wikidata[1] or the Linked Open Data Cloud.[2]

Mirroring the distributed nature of the data landscape, data seekers make use of a variety of resources and strategies to find data. Individuals look for data directly in the above sources, but they also discover data online via academic literature, general search engines, code repositories, forums, and data catalogues which bring together data from multiple places on the web. Nor are all of the sources which people turn to digital.

Four sources for discovering data emerge as being particularly relevant across studies on human-centered data discovery: search engines (both general web search engines and those designed specifically for data); data portals and repositories; academic literature, and people, as data seekers make use of social connections and their own personal networks to discover and understand data.

---

[1] https://www.wikidata.org.

[2] https://lod-cloud.net.

In this chapter, we look at these four sources in more detail. When applicable, we provide an overview of the technical structure of these sources in order to demonstrate the constraints under which data discovery can be understood. We then review what we know about how data seekers interact with these sources to find data. As in Chap. 3, we end by synthesizing findings across the reviewed studies to draw four conclusions about how people discover data.

4.2   Search Engines

4.2.1   Technical Overview

4.2.1.1   General Search Engines

General web search engines, e.g. Google, retrieve results through a process of *crawling, indexing, ranking* and serving results to a searcher. Bots crawl the web to look for new content and web pages, following links from discovered pages to other pages. Web pages are then analyzed for patterns to determine information about their content and are stored in an index. When an individual performs a search over this index, the search results are delivered in a ranked list, whose order is influenced by the number of links between pages (Page et al. 1998).

These steps—crawling, indexing, and ranking—are still at play as individuals search for data using general search engines. Problems arise, however, as algorithms which are developed to work on unstructured documents do not work well on structured data (Cafarella et al. 2008); keyword searches which are common in search engines are also known to be less effective (Lopez-Veyna et al. 2012). It is worth noting, however, that there is a vast amount of research which explores the searching and ranking of structured documents, both in the information retrieval and in the semantic web community (Balog et al. 2010; Dai et al. 2017; Bron et al. 2010).

Unlike when searching for other objects on the web, search engines rarely search through the data themselves, searching metadata or descriptions of data instead in order to determine the relevance of a query. This poses another problem, as these metadata are inconsistent in quality, if they are even present (Atz 2014; Brickley et al. 2019), which limits the accuracy of the search. Page rank also has links between online documents at its core; links between data and other documents, or other digital data, do not exist at the same extent as links between web pages, creating another challenge for data search (Cafarella et al. 2011; Ben Ellefi 2016).

4.2.1.2   Development of Data-Specific Search Engines

Recently, search engines geared specifically toward data retrieval have been developed in order to address these challenges. Google Dataset Search,[3] which debuted in 2018, crawls

---

[3] https://datasetsearch.research.google.com/.

the web for data. Retrieved data are described with the *dataset* class in the schema.org[4] markup language or are described using DCAT,[5] an RDF vocabulary designed to aggregate metadata across data catalogs. Dataset Search collects and enriches this metadata by linking it to other resources, identifying replicas, and reconciling the metadata to the Google knowledge graph. An index of enriched metadata is made for each dataset; the index can then be queried via keywords or CQL expressions (McCallum 2006).

Academic organizations and publishers are also developing general search engines for data. In 2016, Elsevier introduced a beta version of DataSearch, a search engine geared toward finding data within academia. DataSearch also relies on descriptions of data provided in metadata fields. Similar to Google Dataset Search, DataSearch has experimented with ways to enrich and expand provided metadata in order to address problems such as vocabulary mismatches between search queries and incomplete or missing metadata (Scerri 2016). DataSearch itself no longer exists as a separate search engine, but has since been integrated into other Elsevier products, i.e. Mendeley Data[6] and Data Monitor.[7]

Data-specific search engines have also been developed for particular disciplinary domains. Such search engines implement crawlers to discover data which are described using bespoke metadata schemas. DataMed, e.g., is a biomedical search engine which utilizes the Data Tags Suite (DATS), a custom metadata schema describing data access and use conditions, to allow a crawler to automatically index data for search (Sansone et al. 2017; Alter et al. 2020). Other domain-specific search engines have been developed within the context of data repositories, i.e. at GESIS, an established social sciences repository in Germany (Hienert 2019). We further discuss the search functionalities of data repositories and portals in Sect. 4.3.

Emerging areas of research investigate how to extract data mentions from the academic literature for indexing (see Lane et al. 2020). Searching over linked open data is another area of research and experimentation as a way to facilitate data search at web scale, but limited to the Web of Linked Data using very specific query types (e.g., Hartig et al. 2009). Much work has also been done in the area of searching linked data in semantically heterogeneous and distributed environments (Hogan et al. 2011; Freitas et al. 2020), where semantic links are used to come up with an estimate of the importance of each dataset and rank search results.

4.2.1.3 Displaying Search Results

Interface design and the presentation of results on a search engine results page (SERP) are critical to provide context for a search and in making decisions about relevancy. Although not specific to searching for data, much work has been done to investigate how factors such as the number of results per page (Kelly and Azzopardi 2015) or the position of a

---

[4] https://schema.org/.

[5] https://www.w3.org/TR/vocab-dcat-2/.

[6] https://data.mendeley.com.

[7] https://www.elsevier.com/solutions/data-monitor.

result in ranked lists or in grid interfaces can influence user behaviors and perceptions of trustworthiness (Kammerer and Gerjets 2010). Although it is common in bespoke search verticals, i.e. image search (Zhang and Rui 2013), video search (Spolaôr et al. 2020; Schoeffmann et al. 2015) or e-commerce (Rowley 2000; Tsagkias et al. 2020), to present results in alternative ways tailored to specific information needs, this is not yet the case in many data-specific search engines.

Short textual summaries, or snippets, which appear on SERPs after the name of a retrieved item, also help users to evaluate and make decisions about the relevancy of search results in web search (Bando et al. 2010). The content of snippets is based on the specifics of a query, which helps to make selection a more effective process (Tombros and Sanderson 1998; White et al. 2003). Although there are initial efforts to replicate this in data search (Au et al. 2016), there is still much work to do to create a user experience similar to that in web search.

Metadata shown on a results page can also be a means of providing context for a user. In specific types of search, i.e. when searching over collections of information such as products, academic publications, or music, metadata are often displayed alongside search results. Such metadata can include information about publishing dates, author names, author affiliations, etc. Marchionini and White make a distinction between overviews, or "surrogates," and metadata, stating that overviews are specifically designed to support people's initial sensemaking efforts before they more fully engage with information (2007).

Metadata can serve a purpose similar to that of overviews, particularly in data search. Metadata accompanying data are crucial in human-data interaction, as they prompt users to download or access the data, to search further, or to engage with data more fully (Koesten 2019). The SERPs of data-specific search engines initially implemented a ranked list design following the so-called "ten blue links paradigm," which presents users with a list of ten search results with corresponding textual snippets and URLs (Hearst 2009). Recently, some search engines, i.e. Google Dataset Search, have begun to incorporate summaries which highlight key metadata elements; this reflects the evolution of the development of data search engines, but also potentially the incorporation of recommendations from research, which are further discussed in Chap. 6.

It is worth noting the limitations of these approaches to data search. Metadata underlies most of these methods; data must be marked up with schema.org or described using a bespoke metadata schema to be crawled. Data which are not described in these ways cannot be found using search engines. Metadata themselves are often incomplete, are provided in unstandardized formats, and are not curated. This affects the discoverability of data, but is also problematic as people make decisions about the relevance and trustworthiness of data based on metadata fields. Data and metadata curation can enhance both data discoverability and potential reuse, but these efforts are time and cost intensive (Pienta et al. 2017).

4.2.2   How Do People Discover Data Using Search Engines?

4.2.2.1   General Search Engines

Google serves as a starting point for many data seekers. People perform Google searches when they don't know where data might live on the web or when looking for data outside of their domain of expertise (Koesten et al. 2017); Google is also used to locate particular data repositories or perform known-item searches for well-known data (Bishop et al. 2019; Friedrich 2020). This is mirrored in log analysis studies conducted by data repositories, which find that the majority of users enter repositories through Google searches, rather than visiting a repository directly (Pienta et al. 2017). Data seekers often go directly from a Google search to the data's landing page within a repository or portal, bypassing a homepage which could provide extra context about the overall collection (ibid.).

This preference to begin data search with Google could be due to a variety of factors. Searchers may find it more efficient to get to a repository through Google, or they may be searching for a repository with a particular focus. It could also be that they do not believe that the search functionality of a portal is good enough to find the data which they need (Koesten et al. 2017). Part of the allure of general search engines could also be their familiarity and ease of use or the ubiquity and freely-available nature of such tools (Gregory et al. 2020a). It could also be a lack of knowledge of the availability of other search tools which are more specific to data (Krämer et al. 2021).

Even though general search engines are frequently used, it does not mean that they always retrieve the desired results. Researchers and data support professionals report having mixed success with Google, perhaps because of a lack of specificity in search options or in the quality of the content (Gregory et al. 2020a). Assumptions about Google can also inform how people search for data with general search engines, as searchers expect the results of web searches to be worse for data than for academic literature (Kern and Mathiak 2015).

Interview studies provide a window into understanding *how* people use general search engines to locate data. In a study conducted by Koesten and colleagues, data professionals report conducting broad keyword searches consisting of few keywords, followed by subsequent filtering of search results. Data seekers also add data-related terms to Google searches, i.e. terms such as "data", "statistics" or words related to data formats (2017).

In a simulated observational study of data seekers, this behavior was also seen, with participants performing keyword searches with data-related terms in both Google and Google Scholar (Krämer et al. 2021). Queries in this study also tended to be short, although participants reported both broadening and narrowing keyword searches by adding or deleting terms. This last finding contrasts with the results from a large search log analysis of four data portals, where eighty percent of search sessions consisted of only one query (Koesten et al. 2017; see also Sect. 4.4.2).

4.2.2.2 Data-Specific Search Engines

There are a lack of studies documenting how people use data-specific search engines. Benjelloun and colleagues analyze the use of Google Dataset Search by looking at a snapshot of the number of unique datasets which were retrieved during a two-week period in May 2020 (2020). 2.1 million unique datasets from 2.6 thousand domains were listed in the top 100 search results during the study period. The topics of the retrieved data were inferred from associated metadata and webpages. The most frequently retrieved data came from the social sciences (24.5%), followed by biology (14.6%) and medicine (10.8%). These percentages do not reflect the topic distribution of the corpus, where 19% of datasets are from the geosciences. The authors hypothesize that this could be a result of an increased interest in biomedical data during the COVID-19 pandemic (ibid.).

Other analyses focus on the success of Google Dataset Search from a more technical perspective. Alrashed and colleagues found that webpages from 61% of internet hosts that provide data using the dataset class of schema.org do not actually describe datasets (2021). To address this, the authors categorized the webpages where the schema.org markup was not reliable and developed a neural network classifier to successfully indicate if a marked-up object is actually a dataset or not.

Studies with more discipline-specific data search engines mirror some of the challenges users face with general search engines in data search. In a usability study of the DataMed search engine, participants reported having difficulty fitting their complex data needs into a simple search box and in understanding the overall scope of the resource to know which data were actually being searched (Dixit et al. 2018).

Although there is a recognized need for systematic user studies (Brickley et al. 2019), the overall lack of work examining how people use data-specific search engines points to the fact that such tools are still relatively young in their technical development and are perhaps not yet widely known.

4.3   Data Portals and Repositories

4.3.1   Technical Overview

4.3.1.1   What Are Data Portals and Data Repositories?

The terms *data portal* and *data repository* are often used interchangeably, although there are some differences between the two. *Data portals* are online platforms which support accessing data by bringing together lists of links to datasets or to other collections of data. Such portals can be specific to particular audiences or they can be more general.

One type of portal which regularly surfaces in the literature on data discovery are open data portals. These portals provide a point of free access to governmental or institutional data which have been cataloged and archived, allowing users to browse and search through the collected data. Initiatives such as the European Data Portal[8] or Neumaier et al.'s portal watch[9] (2016) build meta-portals that crawl the web to offer integrated access to data drawn from multiple *data repositories* and smaller portals on the web. These data are manually curated, and their schemas and attributes are mapped with each other to make them discoverable within the portal (as described in Brickley et al. 2019).

Data repositories also provide access to data and data collections; this term is frequently used in discussions about data within academia and research. As with portals, data repositories can be targeted toward particular disciplinary domains and data types, i.e. ICPSR[10] for the social sciences, GenBank[11] for genetic sequence data, or Inspire-HEP[12] for high energy physics. Repositories can also be general or multidisciplinary in scope, i.e. Zenodo[13] or Figshare.[14] A series of standards have been developed to certify and regulate data repositories, including the CoreTrustSeal[15] (Dillo and de Leeuw 2018) and DIN31644/NESTOR (Deutsches Institut für Normung 2012). More recently, the TRUST principles have been proposed as a framework for creating "Trustworthy Digital Repositories" (Lin et al. 2020). Such repositories are characterized by having reliable infrastructures, long term governance plans, and policies supporting community-based practices for achieving sustainability and trustworthiness.

4.3.1.2 How Do Data Portals and Repositories Work?

Data portals and repositories have similar technical features. As with search engines, data search within portals and repositories relies on metadata. One of the most popular pieces of software used in open data portals is the Comprehensive Knowledge Archive Network, CKAN,[16] which indexes the metadata provided by data depositors and publishers. As data and metadata are deposited by individual data owners, there is therefore not the same need to discover data "in the wild" as there is for search engines.

Other commercial repositories and portals, i.e. Figshare, and those using Dataverse,[17] also operate in similar ways, providing keyword or faceted search over the metadata of a pool of data. As with discipline-specific search engines, repositories develop and implement unique metadata schemas specific to particular data and communities, i.e. for

---

[8] https://data.europa.eu/en.

[9] https://data.wu.ac.at/portalwatch/.

[10] https://www.icpsr.umich.edu/.

[11] https://www.ncbi.nlm.nih.gov/genbank/.

[12] https://inspirehep.net/.

[13] https://zenodo.org/.

[14] https://figshare.com/.

[15] https://www.coretrustseal.org/.

[16] https://ckan.org/.

[17] https://dataverse.org/.

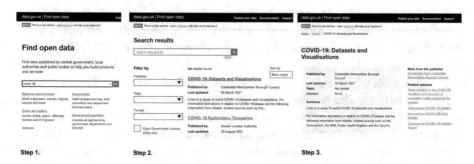

Fig. 4.1 Search steps in the UK governmental data portal. Adapted from Koesten (2019)

ecological data (Michener et al. 1997). They also expand and adopt more general meta-data schemas, i.e. DublinCore or DataCite,[18] to describe data and make them searchable (Strecker 2021).

Data repositories sometimes also facilitate searching for data in relation to other objects. The GESIS repository for social science data integrates the search for literature and data at varying levels of granularity (Hienert et al. 2019). Data seekers can search for data, for literature, or perform more detailed searches for particular variables or survey questions; items within the repository are linked so that users can both discover related items and understand them in context.

4.3.1.3   Search Interfaces

Many data portals and more general data repositories incorporate search interfaces similar to that shown in Fig. 4.1, the interface of the UK governmental data portal.[19] In such portals, data search begins with users entering a query in a simple search box (Step 1); they are then presented with a results list similar to that in general web search (Step 2), where the interface also includes a query bar and series of facets to further filter the results. Upon selecting a result for further exploration, data seekers arrive on a preview page which contains additional information describing the data and a means for downloading the data (Step 3).

This page also usually displays some metadata, including information about format, the affiliated organization, data, license, and topic. In some instances, the landing page also provides a preview or visualization of the data.

Discipline-specific repositories and portals also adopt more bespoke interfaces, including ways to search and browse geographic data or to specify other data characteristics. In a repository for oceanographic data, e.g., users can specify requirements relating to pH levels, oxygen amounts, and geographic area using a combination of search boxes, selections and map interfaces (Megler and Maier 2012). Map-based interfaces and visualizations are

---

[18] https://datacite.org/.
[19] https://www.data.gov.uk/.

commonly used within geospatial or observational data portals and repositories, which provide tools to visually explore data (Dow et al. 2015; Gonzalez et al. 2010).

Implemented in this way, data search within repositories and portals has many of the same limitations discussed in Sect. 4.2.1, which are due in large part to the quality of metadata and to a lack of capabilities for matching keyword-based queries to the data.

4.3.2   How Do People Discover Data in Data Portals and Repositories?

Queries in data portals differ from queries issued in general web search for other types of documents in both their length and structure. In an initial query log analysis of four data portals, supported by a more detailed follow-up log analysis, Kacprzak and colleagues found that queries in data portals are short in length and suggest that data search tends to be exploratory in nature, as data seekers begin with short keyword searches which are subsequently narrowed using filters available from the results list (2017, 2019).

In the follow-up study, the authors propose that governmental data search might be a work-related task, as the majority of queries were issued during working hours from desktop computers. Data seekers entered the portal primarily via Google, although the exact query formulation used in Google was not recorded. Queries within the portal also included geospatial and temporal information, as well as short, data-specific keywords about format and file type (Kacprzak et al. 2019), a finding mirrored in reported data search behaviors within general search engines (Sect. 4.2.2).

The same authors combined their search log analysis with in-depth interviews with data professionals who seek data (Koesten et al. 2017). Data seekers in the study struggled to find relevant data; they iteratively searched, performing multiple searches and refinements. They also used short keyword queries, perhaps because they were not confident that the specificities of their data needs could be met via the initial query box. An ongoing survey conducted by Research Data Austria also found that people searched for data in an exploratory fashion, but that data seekers also used the portal to find more general information, or to locate known items (Wu et al. 2021).

Xiao and colleagues (2020a) combined a transaction log analysis with web context mining to explore data seeking within three data portals in the United States. As in other log analyses, the majority of data seekers arrived at one portal via Google searches; in contrast to previous studies, however, users of the other two portals entered the portals directly or via referrals from other webpages (ibid.). Across the three analyzed portals, data seekers browsed more often than searching. When searching, they refined initial queries by adding geospatial and temporal keywords as well as terms related to data format and file type, perhaps reflecting a lack of corresponding filters for this type of information.

A study of two years of search and interaction logs of the European Data Portal[20] supports the preference for browsing within data portals, highlighting the importance of relevant filters in successful data searches (Ibáñez et al. 2020). Unlike other log analyses, this study provides more detailed information about the relation of initial Google searches to data seeking within portals. While general search engines were often the starting point for people to enter the portal, searches which began with search engines were less successful than those which began within the portal itself. Queries issued via general web search engines to arrive at the portal also tended to be longer, containing more temporal or geospatial keywords than portal queries. However, within-portal queries, which often just consisted of a single word, also contained geospatial terms, i.e. the name of a country, suggesting that users might use the search box itself as a means of filtering results (ibid.).

In contrast to open portals, details about how users search within data repositories and archives are sparse (Borgman et al. 2015), perhaps because many repository logs are not made publicly available. This is particularly true for general data repositories, such as Zenodo or Figshare. This could indicate that search features within these more general sources are underutilized or are not prioritized for development, perhaps because data seekers arrive at such repositories via web search engines.

Studies which have been conducted within more discipline-specific repositories reflect similar search patterns as those presented above, with a tendency towards short keywords which include geospatial and temporal terms. In an analysis of ICPSR,[21] Pienta and colleagues further found that two-thirds of the 500 most frequently searched terms within the repository relate to the *subject* or topic of data (Pienta et al. 2017), suggesting that repositories should focus their efforts on enriching and standardizing subject descriptions of data.

An interview study conducted with the consumers, producers and archivists at a data archive also uncovered challenges in using geospatial keywords to locate data, as the naming conventions for geospatial data within the local area were complex, various and not always in English (Borgman et al. 2019). Study participants also reported a need to be able to browse by subject and have search and browsing features specifically tailored toward geospatial data.

The data which people need can be distributed across numerous data repositories; data from these repositories do not always feed into larger data portals. Data seekers must first discover a relevant repository, and then invest significant time and energy becoming familiar with each individual search environment (Ames et al. 2012; Beran et al. 2009). Even then, the general search interfaces and data descriptions within repositories and databases may not meet their specific search criteria and data needs (Zimmerman 2007).

---

[20] https://data.europa.eu/en.
[21] https://www.icpsr.umich.edu/.

4.4 Academic Literature

4.4.1 Technical Infrastructure

While it is beyond the scope of this chapter to extensively review the technical underpinnings of document and literature search, a few points are worth mentioning within the context of data discovery. The first relates to the structure of academic literature, particularly regarding the presentation of data and data citations. The second relates to how academic literature is typically made searchable and retrievable, with special attention paid to the role of citation searching.

4.4.1.1 Structure of Academic Literature

Data are a key part of research publications, although they are not always directly accessible or available in their entirety. Data are visualized and recorded in figures, graphs, charts and images which are embedded within articles; they are also listed in tables or within the body of the text itself. Data, visualizations, tables, and instruments, i.e. survey questionnaires, can also be attached as supplementary information or they can be stored in a separate location, i.e. on a personal webpage or in a repository, and be linked to an article (Park et al. 2018).

Publishers are increasingly requiring data availability statements, in which authors describe if and how the data supporting a publication can be accessed. Many of these statements state that "data are available upon request," but this is not a guarantee that authors will in fact share their data if they are contacted. In an analysis of articles with data availability statements in psychology, for example, 73% requests for data went unanswered (Wicherts et al. 2006). Data linked from publications to personal web pages also suffer from link rot and are often not available in a sustainable way (Pepe et al. 2014).

Organizations such as DataCite[22] and Make Data Count[23] work to develop the technical and social infrastructure necessary to facilitate formal data citations. This involves assigning persistent identifiers to data, developing metadata schemas to capture citation relationships between articles and data, and working to ensure that adequate metadata are provided. Different entities, including publishers and professional associations, also issue guidance on recommended formats for data citation. While there is not yet agreement about the exact format of a data citation, recommendations convene on including a digital object identifier (DOI) or other persistent identifier to reference data, in part to ease data discoverability and accessibility.

Citing data within the academic literature is not yet a common practice. Certain disciplines, such as biomedical fields, cite data more regularly than others, although even within biomedicine, citation practices vary and the use of DOIs is not prevalent (Peters et al. 2016). Many data which are published in data repositories and portals do not have

---

[22] https://datacite.org/.

[23] https://makedatacount.org/.

complete metadata, particularly regarding subject information. In a 2021 study of nearly nine million datasets in the DataCite corpus, only 6% of data had subject information; only 1% of the data in the corpus had been cited within the academic literature (Ninkov et al. 2021).

4.4.1.2 Search and Retrieval Systems

Within the bibliographic database environment, academic literature is organized and made searchable via records containing keywords and metadata describing documents and their sources, including information about how to access the data. These records serve as surrogates for the documents themselves. Search is performed over the bibliographic records via search interfaces in individual databases or via general discovery tools which search across library collections.

Many initial search interfaces in databases, online public access catalogs, and discovery tools include a basic query box, similar to that of the search interfaces in data portals presented in Sect. 4.3.1. A variety of statistical and semantic techniques are used to compute the relevance of search terms and documents, which are then displayed on a results page where further filtering is possible. The quality of the metadata available in bibliographic databases is generally superior to that available for data in data portals and repositories. Bibliographic databases also have more specialized filters and advanced search options which build on well-curated metadata.

General web search engines, and also more specialized search verticals, i.e. Google Scholar, are other key systems used to discover academic literature. Web search works as described in Sect. 4.2.1, although the full text of documents are indexed and searched in literature search, as opposed to just the metadata. Building on the popularity of commercial citation databases within academia, Google Scholar also has the additional feature of forward citation searching, allowing searchers to see which other publications have cited a particular work.

Commercial citation databases, i.e. Web of Science from Clarivate Analytics and Elsevier's Scopus database, facilitate both forward citation searching and tracing references to earlier works from publication bibliographies, so-called backward citation searching (Araujo et al. 2021). Forward and backward citation searching is a key strategy employed within academic literature search.

The Data Citation Index (DCI), now owned by Clarivate Analytics, was created in 2012 to capture data citations in the academic literature The DCI indexes data from selected repositories, which are chosen according to specific criteria including subject, editorial content, geographic origin and scope (Clarivate 2022). Data are classified into three hierarchical categories—repositories, data studies and datasets; metadata describing lower level categories, i.e. datasets, is derived from metadata available for higher level categories, i.e. repositories (Force and Robinson 2014). Data within the DCI are searchable using interfaces similar to those in other bibliographic databases. Data records, or metadata, can have links to literature where the data have been cited or include references to other resources which informed the data.

DataCite, founded in 2009, is a non-profit organization which links research, including data and documents, by assigning DOIs and capturing existing relationship types between them. Repositories who become members of DataCite describe their data using the DataCite Metadata Schema in the process of having DOIs assigned. This metadata schema is general, designed to facilitate discovery and citation, rather than to replace more detailed community-specific schemas. Only six metadata fields are required to describe data in the DataCite schema: *identifier, creator, title, publisher, publication year* and *resource type*. Additional fields for relation type indicate relations between two documents, i.e. "cited by" and "cites". DataCite provides access to its corpus via an API and the DataCite Commons[24] interface, where users can search by works (i.e. data or literature), people and organizations to discover links between these different types of objects, including citation relationships.

4.4.2 How Do People Discover Data Using Academic Literature?

The academic literature is a key source for discovering data for researchers across disciplines. In a survey on data discovery practices with nearly 1700 respondents, only 6% of respondents reported never making use of the academic literature as a source for locating data (Gregory et al. 2020a). This reliance on the academic literature to locate data has also been documented in other studies within specific disciplines, such as geography, archaeology, and the environmental sciences (Borgman et al. 2005; Faniel and Yakel 2017; Schmidt et al. 2016; Zimmerman 2007, 2008).

Researchers use academic literature to find data in perhaps unexpected ways. They pluck data from figures, tables and graphs (Pepe et al. 2014) and discover data serendipitously while reading or searching for literature for other purposes. Researchers also explicitly perform literature searches with the goal of finding data (Gregory et al. 2020a).

This highlights the fact that data discovery does not always exist as a separate practice from literature searching. In the previously mentioned survey, 30% of respondents reported *no difference* in how they find literature and how they find data. The authors also suggest that practices of data discovery and literature discovery are more integrated with each other in fields which have data repositories which are closely linked with systems searching the academic literature within a field, such as the biomedical sciences or physics (Gregory et al. 2020a).

Following citations from the literature to data is a key discovery strategy. Nearly 80% of researchers who reported using the literature as a data source in the Gregory et al. survey indicated that they trace citations from the literature to data. Other studies have also reported that researchers follow references to data from the literature, to locate both

---

[24] https://commons.datacite.org/.

known data and unknown data (Faniel et al. 2013; Friedrich 2020; Kern and Mathiak 2015; Zimmerman 2007).

The lack of formal citations to data in the literature and the prevalence of link rot in more informal data references, reported in Sect. 4.4.1 indicates a mismatch between using citation chaining as a search strategy and the available infrastructure and referencing practices which would enable successful citation searches.

There are also a lack of studies investigating how people use the Data Citation Index and the DataCite API to actually locate data and follow citation relationships. The majority of studies conducted thus far analyze the content of each source, i.e. the disciplinary distribution of data, data types and data supplies, particularly with an eye to assessing the suitability of these sources for bibliometric analyses (Robinson-Garcia et al. 2017; Robinson-Garcia et al. 2016; Peters et al. 2016).

Kern and Mathiak further suggest that searching for data is a more complex process than literature searching. In two user studies with quantitative social scientists, they found that metadata quality was more important when searching and evaluating data, and convenience was more important in literature search (2015). They suggest that researchers are willing to invest more extra effort and time when searching for data than when seeking academic literature (ibid.).

Hienert and colleagues tracked the search trajectories of data seekers when using a search engine for both data and related documents (2019). Most of the links between searched objects were between data and publications, and vice versa, showing that the literature is not only important in discovering data, but also in providing context to evaluate and understand data once they are found (Friedrich 2020; see Chap. 5).

4.5 Social Connections

Data seekers across studies consult with other people to discover, access and make sense of data. These exchanges occur in person, but they are also mediated by a variety of technologies, occurring through email exchanges, discussions on platforms such as Kaggle, Stack Overflow or GitHub, or on social media. In this section, we forgo a technical description of the many technologies which are embedded within social practice, focusing instead on how people interact with others to discover data.

4.5.1 How Do People Discover Data Through Social Connections?

Personal exchanges are valuable sources for both discovering and accessing data. In some cases, individuals identify data or a data repository and then contact the data owner or repository manager for access. For instance in astronomy where researchers working on the Sloan Digital Sky Survey browse personal websites to locate data and then contact research groups directly (Sands et al. 2012). These types of personal exchanges can be a multi-step process where data seekers converse with data providers to ascertain the suitability of data, before those data are gathered, processed and delivered (Wallis et al. 2013), or where data seekers draft emails, write letters, and make telephone calls to obtain access to data which they first discovered in articles (Wallis et al. 2007).

Personal communications are perhaps especially important in tracking down small, specialized datasets, which are more difficult to locate than gold-standard data which are well known (Sands et al. 2012; Swan and Brown 2008). Even in disciplinary fields which have well developed data and information infrastructures, data seekers rely on personal networks to identify and access data (Sands et al. 2012; Gregory et al. 2020a).

Researchers engage in both serendipitous and more strategic interactions with co-workers, supervisors, collaborators, data authors and former employees to discover data (Gregory et al. 2020a; Yoon et al. 2014). Early career researchers in particular rely on recommendations from their advisors and tips from colleagues (Kriesberg et al. 2013; Faniel and Yakel 2017). Discovering needed data can be difficult when data seekers do not know whom to contact for this advice (Curty 2016).

Professional experience also influences how individuals make use of social connections to discover data. Individuals seeking survey data who have lower levels of expertise consult with supervisors and data librarians to find data; researchers with higher levels of expertise engage more with peers and attend conferences to locate data (Friedrich 2020). Friedrich also investigates the role of data support librarians and specialists in the data discovery process (ibid.). Most often, researchers consult with support staff in problematic cases, i.e. when data cannot be found or accessed easily or when the data or documentation have suspected errors.

Similarly, Koesten et al. and Kacparzak et al. suggest that the majority of their study participants made use of social interactions to find data due to limitations in the search functionality of data portals and web search engines for finding data (Koesten et al. 2017; Kacparzak 2019). People also consult with experts if needed data are not digital, i.e. if data are hidden in museum archives or in governmental offices, or if data are not made searchable (Gregory et al. 2019). Search difficulties are not the only reason why people make use of social interactions. Speaking with other people may provide affordances which are not supported in online search systems, such as developing social and professional networks, forming new research collaborations (Gregory et al. 2020b), or collaboratively understanding data (Koesten et al. 2021).

4.6   Conclusions About Discovering Data

In this chapter, we reviewed four common sources used to discover data, from both a technical and interaction perspective: search engines, data portals and repositories, academic literature, and social connections. Based on this synthesis, we propose four conclusions about how people discover data in these sources.

Many Searches Start at Google

Many people begin searching for data using a general web search engine. They search using short keyword searches to discover new data, to find and access data within portals and repositories, to locate repositories themselves, and to find known datasets. Although keyword searches are common, they are not as effective for data as for other types of web search.

Basic Search Boxes Alone Do not Meet the Specificity of Data Needs

Basic query boxes offer a starting point for further exploratory data searching, rather than providing a direct route to needed data. Keyword searches, which commonly include geospatial, temporal and data-related terms, are followed by browsing using filters, if they are available. Data seekers need high quality metadata to support both browsing and searching; this metadata is often not present or is incomplete.

Search is but One Way of Discovering Data

Data seekers make use of a variety of sources and strategies to locate data, not all of which involve a search engine or data portal. Data search is linked with other types of search, i.e. literature search; data seekers also make use of social exchanges to discover, access and understand data and to solve data-related problems.

Infrastructural Limitations and System Design Shape Data Discovery

Data are not always findable or reusable. Limitations in infrastructure, such as unstandardized and incomplete metadata, uncurated datasets, a lack of data citations or incomplete collections determine what data can be found, and thus reused. Data discovery systems have also been developed based on common practice in web and literature search, particularly regarding interface design and a reliance on keyword search; this in turn shapes the practices of individuals searching for data.

Data Evaluation and Sensemaking

5

This chapter discusses how individuals evaluate and make sense of data which they discover. Evaluating data for potential use includes drawing on different types of information *from* and *about* data (Sect. 5.1). Once selected, individuals go "*in*" the data, as they try to make sense of the data themselves (Sect. 5.2). We emphasize that this process of evaluation and sensemaking is not sequential. It includes different cycles of evaluating, selecting and trying to explore data in more depth. This can in turn lead to refining one's selection criteria and going back to evaluating and selecting different data sources to start the process again. The chapter ends by drawing three conclusions (Sect. 5.3) about how people evaluate and make sense of data for reuse.

5.1 Selecting Data: Common Evaluation Criteria

A key moment in data discovery is when a person finds data and is left with the decision of whether or not to "click" on a link or to use the data which are found. More often than not, this turns out to be a complex decision, involving iteratively exploring different aspects of a dataset. It can also include performing exploratory analyses to evaluate the data's fitness for use.

But what are the different factors people use to evaluate data? As noted in earlier chapters, these aspects are inherently dependent on specific data needs. However, synthesizing across different data needs and literatures, we summarize important characteristics that people use to evaluate data in this section under the term "evaluation criteria".

We split our discussion of evaluation criteria into two categories: information *from the data* and information *about the data*. The first category, Information *from the data*, refers to information which is contained within the data themselves. Section 5.1.1, which is based on the work of Koesten et al. (2017, 2020), Koesten (2019), further discusses

K. Gregory and L. Koesten, *Human-Centered Data Discovery*, Synthesis Lectures on Information Concepts, Retrieval, and Services, https://doi.org/10.1007/978-3-031-18223-5_5

this type of criteria in three high-level categories: *relevance, usability* and *quality* criteria. While these are common categories linked to information objects in general (Bales and Wang 2005), literature shows that the practice of how to assess these dimensions can be specific to data (Wynholds et al. 2011; Kern and Mathiak 2015).

We then discuss the contextual information needed *about* data for evaluation. Such information is located "outside" of the data, and is contained in metadata, other forms of documentation, or other information structures (see Sect. 5.1.2). These two categories are not mutually exclusive, and the boundary between them is at times indistinct. Certain criteria, i.e. information about data collection methods, can potentially be derived both from the data themselves and from data descriptions. Nevertheless, the differentiation we draw between the two categories can be helpful in providing an overview and nuancing evaluation practices.

5.1.1 Information from Data

5.1.1.1 Relevance Criteria

For literature, the concept of relevance changes depending on authors and time. How people assess the relevance of documents has been studied extensively (Barry 1994; Park 1993; Bales and Wang 2005). Aiming to understand the criteria people use to determine relevance in information retrieval systems, Bales and Wang (2005) identified 133 criteria in a review; in the same spirit, the notion of a whole "system of relevances" has been discussed (Saracevic 1997).

Relevance for data can also be seen as a broad concept which refers to whether the content of data is considered applicable to a particular task. A key question for data seekers is whether data are about the correct *topic*. A related factor is the temporal and geographical scope of data, i.e. if data have information about the right time and the right area, at the needed level of detail. The original purpose of the data and external requirements or norms that might have influenced data creation are also mentioned in data search specific studies (i.e. Koesten et al. 2017) to impact people's relevance judgements. This type of information is essentially "baked into" or embedded within data, as choices about data collection methods and documentation, for example, are visible within the data themselves.

5.1.1.2 Usability Criteria

Usability includes the suitability of data with regards to different types of practical implications. In order to judge the data's usability for a given task, people consider factors such as format, size, documentation, language (e.g., used in headers or for string values), comparability (e.g., identifiers, units of measurement), references to connected sources, or ease of access.

5.1.1.3 Quality Criteria

For purposes of data evaluation, quality criteria can be thought of as anything that individuals use to judge data's condition or standard for a task. Quality is a multidimensional concept that can also include judgements about data completeness, accuracy, timeliness, methodological choices in data creation, or missing values.

Specific data types afford certain evaluation criteria. The dimensions that constitute data quality are to an extent implied in data structures. Some of these dimensions pertain to the data type or format and some to data collection methods. For example, individuals working with static CSV files check them for consistency, missing values, and machine readability to determine their quality (Koesten et al. 2020). Others check dynamic sensor data for measurement error, plausibility, timeliness, and efficiency (Karkouch et al. 2016). People evaluating qualitative data may try to understand a coding scheme, categorisation efforts or sampling criteria (Bishop and Kuula-Luumi 2017; Poth 2019).

The literature about data quality research proposes quality metrics for different types of data, including tabular data and linked data (Batini et al. 2009; Zaveri et al. 2016) but also for sensor data, interview data and more. Such metrics include information about consistency, completeness, or interoperability. However, considering the common data discovery sources described in Chap. 4, these metrics may not be readily available to data seekers, due to technical limitations or the fact that they might only be definable for specific tasks. Individuals therefore may need to perform extensive research and data exploration to be able to judge some of these dimensions.

5.1.1.4 Other Related Work

Other studies specifically investigate evaluation criteria in data discovery in different contexts. In a study focusing on research data, Gregory et al. found information about data collection conditions and methodology to be highly important, next to information about data processing and handling as well as the topical relevance of the datasets (Gregory et al. 2020a). Other relevant factors included the ease (or difficulty) of accessing data, timeliness, and the reputation of the source of the data (e.g. the repository or journal) as a means to establish trust. These findings can be mapped to our broad categories of relevance, usability and quality, but they also explicitly include social factors, i.e. the reputation of the data source or of the data creator.

Using an approach based on textual summaries, Koesten et al. identify nine key questions to consider when describing data for others to evaluate (2020). Koesten, Gregory and colleagues expand this work by using verbal summaries of data and focusing on the differences between data within and outside people's areas of expertise (2021). Phillips and Smit draw on expert consultations to identify how to extend this type of work by focusing on research data, also emphasizing that the context of data creation is key for evaluation (2021).

Evaluation criteria are context-dependent (Freund 2013) and are based on broader data needs, including the type of *task* for which data are sought (see Chap. 3). Having access to a range of potentially useful evaluation criteria is key to being able to make informed decisions during the data discovery process. This is highly dependent on both existing metadata as well as the functionalities of search interfaces, as it is impossible to make judgements based on information that is not available (Koesten et al. 2021).

Data do not become usable simply through availability; users must have the necessary information in order to make informed decisions (Koesten and Simperl 2021). The thought that *one person's signal can be another person's noise* (Borgman 2015) needs to be embedded in the design of data search systems to cater to the large variety of different data needs and to support varied evaluation criteria.

As an example, in a study by Koesten et al. a participant described a dataset about the occupancy of local car parks (2017). Due to malfunctioning sensors, the car park could at times appear to be over full. As this was not actually possible, values of more than 100% car park occupancy were deleted from the dataset during data cleaning to match the needs of the car park users. However, if another person wanted to use the data to understand the reliability of the sensors themselves, all of the relevant data would be missing.

Another example can be found when thinking about using the size of data as a criteria for evaluation. Data which are large in size can be a positive quality factor for machine learning datasets, but size can be a negative factor if it means that someone cannot handle a large dataset with standard tools and configurations or easily share it with collaborators. These cases exemplify how different data needs and evaluation criteria impact which data are considered to be useful in different situations.

5.1.2 Information About Data

Not all information needed to evaluate data can be found within the data; some is found "outside" of the data. While we have discussed the importance of metadata throughout this lecture, here we also describe "contextual information" that often goes beyond what is commonly available as metadata.

Necessary contextual information is influenced by factors such as the environment data are situated in and the connected epistemic cultures of different communities (Gregory 2021a). Existing work has investigated contextual information needs in different settings, often focusing on research data (Gregory et al. 2020a; Koesten et al. 2021; Faniel et al. 2019), including studies of specific research disciplines, such as astronomy (Sands et al. 2012), earth and environmental sciences (Zimmerman 2008), biomedicine (Federer 2019), archaeology (Faniel et al. 2013), and the social sciences (Krämer et al. 2021). Recent work, focusing on but not limited to machine learning datasets, has also aimed to create practical guidelines for capturing contextual information (Gebru et al. 2021; Holland et al. 2018; Koesten et al. 2020).

Looking at literature on data reuse, it becomes clear that evaluation criteria are determined by information about the data themselves and about their context; the boundaries of these terms are not clear cut. Here we give a brief overview about types of contextual information and how they are discussed in research.

Faniel and colleagues describe three broad contextual information categories for data based on their work with researchers in three disciplinary domains (2019). These categories include: (i) information about data production; (ii) information about the repository where the data are published; and (iii) information about data reuse. Examples of contextual information in these categories include documentation about instruments, methodologies, research questions, and observational conditions. It can also include information describing the meaning of missing values, of specific abbreviations, or guidance on how to interpret categories. Details about methodological questions of data creation, collection, processing, storage and maintenance, including information about provenance and data versioning, are one of the most commonly mentioned contextual information types. Synthesizing work on the topic, Koesten, et al. compile 40 concrete recommendations from literature for data documentation with the aim of reuse (2020).

Contextual information comes in different forms. It can be provided explicitly with the data, for instance in the form of enriched metadata or codebooks. It can also be provided via a multitude of formal and informal information structures that people use to capture and preserve context around data. Such information structures include supplementary files, notebooks, emails, figures, slides, audio files, images, and messages in private or shared discussion forums (Koesten et al. 2021; Faniel and Yakel 2017; Borgman 2015). These resources are created to avoid losing the meaning of data and to enable reuse, both for the original data producer at a later time, as well as for other potential users.

The nature of this documentation can be different depending on the audience. Data may be documented differently for an individual's own future use, as compared to the use by another person close to the data (e.g. individuals working on the same team or research group) or by people within the same disciplinary domain. The necessary contextual information, the level of detail, and the style of language may vary according to predicted data uses or levels of expertise (Gregory et al. 2021).

Understanding evaluation criteria is a step towards supporting successful data discovery and should inform interface design and the architecture of data discovery systems. As we discuss in the following section, evaluating data for use is inextricably linked with making sense of data.

5.2   Being "In" the Data, or Data-Centric Sensemaking

Using data beyond the context of their original creation can be challenging. People need information about the data to be able to judge them, to know what they can be used for and how they can be understood. Section 5.1 illustrates different types of evaluation criteria which people apply in order to select data for use. Implicit in selecting data is the ability to make sense of data, but what does sensemaking encompass and how is it unique for data, if at all?

The concept of sensemaking has been studied in different disciplines, including psychology (e.g. Klein et al. 2006a, b), specifically decision making (e.g. Klein et al. 1993; Malakis and Kontogiannis 2013) and organizational behavior (e.g. see Maitlis and Christianson 2014 for a review). Other disciplines include educational research (see Odden and Russ 2019 for a review), information seeking (Dervin 1997; Marchionini and White 2007; Savolainen 1993), and human computer interaction (HCI) (e.g. Russell et al. 1993), to name a few.

While sensemaking is an elusive term used in different contexts, we focus here on sensemaking as discussed in information science and HCI to refer to the process of constructing meaning from information (Blandford and Attfield 2010). Sensemaking in these disciplines is recognized as being an iterative process that involves linking different pieces of information into a single conceptual representation (Hearst 2009; Russell 2003).

In their data frame theory, Klein and colleagues define sensemaking as "the deliberate effort to understand events" (2007). In this theory, the definition of data is broad, including different types of stimuli or information that individuals are exposed to. Klein et al. emphasize how the perspective (or frame) of the data consumer shapes the data in terms of how they are perceived, interpreted and even acquired. Through engaging with data, pre-existing frames either get reinforced or change, because the data do not fit, which is where sensemaking happens. Similarly, Koesten, Gregory and colleagues also describe the discovery of *strange things* in the data, such as outliers, errors, missing data, and inconsistencies, as entry points to making sense of data, and as incentives for individuals to explore the data further (2021).

Other authors propose that sensemaking allows individuals to create rational accounts of the world which enable action (Maitlis 2005). Mirroring this perspective, sensemaking is also described as the active processing of information to achieve understanding. Building on the work of Russel (1993), Pirolli and Card explore the sensemaking processes of data analysts as the *iterative development of representational schemas that best fit the evidence and provide a basis for understanding data* (2005).

Koesten, Gregory and colleagues specifically investigate the **sensemaking activities of researchers** in encounters with data in light of potential data reuse (2021). They use an approach to understand sensemaking through summarization that involves observations, semi-structured interviews and a think-aloud task, both with data their participants knew well, as well as data that were new to them. The authors bring together sensemaking and associated evaluation criteria for data, identifying three clusters of activities involved in initial data-centric sensemaking: inspecting, engaging with content, and placing data in different contexts.

The first of these clusters, *inspecting*, involves understanding the general shape of the data. In this phase, people get a broad overview of the data's content, including the topic, size, structure, meaning of the key variables, and the amount of missing values. This is done through activities which include scanning or scrolling through the entire dataset.

The second cluster, *engaging*, includes a deeper examination of the data. Here, people gauge the data's trustworthiness by checking for errors, or for values that look strange or don't fit, in order to see if any parts of the data are missing or do not match expectations. People question the meaning of any identified uncertainties in the context of their planned purposes for using the data. The composition of variables and their relationships to other parts of the data or to contextual information, such as categorization activities during data creation or derived values, are investigated in detail during the *engagement* phase.

In the final cluster, *placing*, data are placed within different contexts to understand their meaning and representativeness. This includes the immediate contexts of data creation, such as detailed study designs, experimental setups or the conditions surrounding data collection. This cluster also includes placing data within disciplinary norms and their social contexts. Taking another step away from the data themselves, *placing* also includes asking questions about the representativeness of data in the world, i.e. regarding temporal and geographic considerations and questions of bias and perspective.

One example drawn from this study were participants' questions about data containing countries as key variables. As illustrated by the following quotes, study participants worked to place data by questioning if the list of included countries was complete. They also questioned the granularity of the data to ascertain if the data were representative of an entire country, or only of certain areas of a country.

Participant 2: It's listing the countries for which data are available, not sure if this is truly all countries we know of...

Participant 7: ...If it was the whole country that was affected or not, affecting the northern, the western, eastern, southern parts?

Participant 24: Was it sampled and then estimated for the whole country? Or is it the exact number of deaths that were got from hospitals and health agencies, for example? So is it a census or is it an estimate?

The authors derive design recommendations from this work aiming to support the identified clusters of activity patterns, which we discuss in more detail in Chap. 6.

Exploratory data analysis (EDA), a term coined by Tukey in 1977, is a process of engaging with data to make sense of them, focusing on the information within the data. It involves summarizing the characteristics of data both statistically as well as visually. In an iterative process, data are explored in different ways, usually without people having a prior understanding of what the data contains, until a "story" emerges (e.g. through patterns, outliers, or comparisons to other data) (Marchionini 2006; Baker et al. 2009).

Studies investigating sensemaking with data in HCI tend to focus on quantitative data and often address the role that **visualization** plays in identifying patterns in data (Furnas and Russell 2005; Kang and Stasko 2012). This focus reflects the emergence of bespoke visual exploration environments (Yalçin et al. 2018; Marchionini et al. 2005). Other work proposes visualization tools to aid in sensemaking activities, such as a visual analytics system tailored for particular groups of data analysts (Stasko et al. 2008) or agile display mechanisms for users accessing government statistics (Marchionini et al. 2005).

Studies on the **comprehension of spreadsheets** have also highlighted the importance of contextual information in sensemaking. Here, an additional complicating factor is undocumented formulas in spreadsheets that can be difficult to decipher for future users, as well as cross sheet referencing (Ragavan et al. 2021, Hermans et al. 2011).

While we have emphasized the impact of data tasks on both evaluation criteria for data as well on the sensemaking process, other factors, such as a data consumer's prior knowledge, skills, experience, motivation etc. also play a role (Kelly 2009; Marchionini 1997). As sensemaking is a cognitive process, these factors are naturally important to consider, as is the fact that data users are constructive actors (Ingwersen and Järvelin 2005) with different dispositions and levels of data literacy. Koesten et al. emphasize the fluid notion of data users, with some being experts for very specific tasks while also being novices for other tasks (2021). Hence we believe the focus on data tasks and needs lends itself more usefully to inform the design of data discovery systems than categorizations of different user types.

More recently, **critical data studies** describe evaluation processes increasingly as collective efforts, due to interpretative layers built into the creation and use of data (Neff et al. 2017). Here emphasis is put onto the sociotechnical environment from which data originates and the resulting implications on what data is and what it can be used for.

Collaborative sensemaking has long been a topic in the Computer Supported Cooperative Work (CSCW) literature (Mahyar and Tory 2014; Goyal and Fussell 2016; Paul and Reddy 2010). Traditionally, using data involves finding and downloading them. With advances in cloud computing, data has become more accessible to online communities, shifting some kinds of data reuse to the online space. This allows users to try out the data on the fly, to talk to others about the data, and to create meaning through these types of interactions. With online data communities there are new spaces to study collaborative and social aspects of sensemaking with data.

The production and consumption of academic knowledge has also been conceptualized as a form of dialogue (e.g. Lillis 2011), where the broader practice of data reuse could itself be seen as a form of collaboration or conversation between data producers and consumers (Koesten et al. 2021). The creation of data often includes collaborative decisions (Mahyar and Tory 2014; Neff et al. 2017), and the use of data is increasingly done in teams (Koesten et al. 2019). Gregory discusses **data communities**, which can form around shared methods for analyzing or creating data, which also points to the collaborative nature of understanding data for reuse (2021).

5.3 Conclusions About Data Evaluation and Sensemaking

People Need Information from and About Data

Data evaluation is dynamic, and involves defining *relevance, usability* and *data quality* according to data needs and types.

While applicable to information needs generally, the specifics of establishing these criteria are data related. This concerns questions of topical relevance and granularity of the data, usability concerns related to aspects of data handling, as well as diverse quality dimensions to establish trust in the data for a specific task.

At the same time, contextual information, both explicit as well as implicit, plays a key role in evaluating data. This includes metadata, other types of structured documentations as well as informal information structures.

People Make Sense of Data by Using Task-Specific Evaluation Criteria and Sensemaking Strategies

Data-centric sensemaking consists of iterative attempts to combine knowledge about data attributes, and different activities to understand data, including social interactions.

Different activity patterns are used to make sense of data, including *inspecting*, *engaging* and *placing* activities. These activities exist on a spectrum from a general look on the data (inspecting), to focused engagement with the data to understand their structure, individual variables and their relationships.

Sensemaking relies on more than just the data, it involves thinking about the context in which the data were created, the purpose, representativeness, and sociotechnical aspects, essentially taking into account the entire environment in which data are situated.

People Make Sense of Data Collaboratively and Through Social Connections

The nature of data work is developing towards more collaborative settings in which both creation, analysis and sensemaking processes are increasingly done in interaction with other people. This influences the design of data discovery and data practice environments.

Recommendations for Data Discovery, Sensemaking and Reuse

6

In this chapter we synthesize and propose recommendations and areas for future work to support human-centered data discovery based on the literature and insights presented in previous chapters. These recommendations fall along three lines: (i) recommendations for the design of search interfaces, which we discuss in the most detail; (ii) recommendations for metadata and data documentation; and (iii) recommendations for supporting community engagement and collaborative work around data. The recommendations are relevant for a variety of stakeholders, including designers of data discovery systems; managers of data portals and repositories; and data creators and data seekers themselves.

6.1 Recommendations for Data Discovery Interfaces (and Systems)

The previous chapters describe what we know from literature about the data discovery practices and data needs of people in different communities. Recommendations of varying levels of specificity have been proposed within the context of the reviewed work. As our focus is on human-centered data discovery and user interaction, we concentrate our recommendations on the design of *data search interfaces,* bearing in mind that the described functionalities need to be supported by backend systems and metadata. The recommendations which we bring together here can be applicable to the data discovery interfaces of data-specific search engines, as well as to search interfaces within data portals and repositories.

Not all of the recommendations which we put forward may be feasible at this point in time; they may also not scale well from a technical point of view or be supported by the necessary metadata. We also recognize that continued research is needed to know how best to support interactions within data search systems. Evidence, such as search

© The Author(s), under exclusive license to Springer Nature Switzerland AG 2022
K. Gregory and L. Koesten, *Human-Centered Data Discovery*, Synthesis Lectures
on Information Concepts, Retrieval, and Services,
https://doi.org/10.1007/978-3-031-18223-5_6

logs about data seeking behaviors is relatively limited. Many existing search solutions, i.e. within open data portals, are often basic, making it difficult to study how data seekers would interact with advanced search systems in the context of real world tasks.

These recommendations are derived from the scientific literature discussed in this lecture, including findings from our own work, and existing best practice guidelines or examples of data discovery systems. They do not aim to be comprehensive, but instead offer a practical summary of the implications of this work in a way which may be useful for different audiences.

Mirroring the human-centered perspective of this lecture, we present our recommendations in this section in the order in which a data seeker might work through a search scenario within a data portal, from (i) a search page to (ii) a results page to (iii) a landing or data preview page. Although we follow this order, we know from earlier chapters that many data seekers may only actually see the landing page itself, having found the data through a general search engine or link from the literature.

6.1.1 Initial Search Page

The initial search page of a data portal, repository or search engine offers an entry point for beginning a data search. As we have seen throughout this lecture, simple keyword searches for data do not offer the same user experience which people are accustomed to in web search. Data needs may be too complex to express in a simple search box, and keyword queries are of limited efficacy. We recommend providing **faceted searching** in addition to keyword queries on initial search pages. This is already partially realized by some interfaces mentioned in earlier chapters; here, we summarize important potential facets and suggest further avenues for exploration.

Determining the most meaningful facets to include should be a continued area of research. This being said, the common evaluation criteria identified in earlier chapters could be used as facets on search pages. Common facets supporting *relevance* determinations include some type of topic categorization and facets related to geographic and temporal scope. Facets for *usability* information include format, size, license/access, and the date of the last update. *Quality* dimensions, as discussed in Chap. 5, could also be considered as facets for inclusion including social signals; search pages could also offer ways to exclude certain data from a search, both through negative keywords or facets.

Throughout the lecture, we have emphasized that *data tasks* are a key part of data discovery. Differentiated search interfaces (as in Baker et al. 2015) or facets could be implemented for particular tasks, data uses, or expertise levels. At the same time we should also consider how to support **serendipitous discovery,** i.e. by highlighting new data or data on particular topics, or recommending data based on past search patterns.

The search page should give an **overview** of how many datasets are available for a certain query or facet to inform personal search strategies, ideally dynamically changing as these search strategies are adjusted. We have seen suggestions that people's mental models for data search are based on web search. Data search systems themselves mimic web search environments, although with limitations which are perhaps not obvious to a user. We cannot assume that data seekers know the constraints of data search, i.e. limited indexing, poor metadata quality and a lack of links between data. Data search systems must translate user behavior with this in mind.

Initial search pages can also offer options beyond basic keyword and faceted searching. Searching by example (Rezig et al. 2021), i.e. by using examples of data or tables as search queries, can also work in certain contexts.

Given the increased use of speech interfaces in other types of search, **voice input** could also be considered in the context of data search. Data needs are often complex and dynamic, and may lend themselves to longer, question-type queries which could be expressed in natural language questions (both as text or as speech input). Query intent classification is also not yet at the level for understanding what data a user may be looking for, or if a combination of datasets may in fact be needed. Making such a classification would require both understanding the context of separate datasets and searching within a particular dataset.

6.1.2 Search Engine Results Page (SERP)

Providing a suitable overview without overwhelming the user requires careful considera- tion of the layout of the SERP. On the one hand, the SERP should support exploring the relevance, usability and quality of data at a more granular level, while on the other hand it should not overwhelm the data seeker. One solution may be to provide different views of the data depending on different types of data tasks.

Following Shneiderman, we recommend designing interfaces that enable zooming in and out of data, allowing users to choose details on demand (1996). While this is not an uncommon approach in visual data exploration tools, general data search interfaces do not commonly support this interaction.

Work throughout this lecture has highlighted the importance of temporal and geospa- tial descriptions of data when determining relevance. We mentioned the possibility of including such information as facets on the initial search page, but we also recommend that ways to explore temporal and geographic characteristics be present at all stages of the search process, at increasing levels of granularity. This might include timeline or map based visualizations. On a SERP, summaries of geographic or temporal elements could be included, as is done already in some data portals; more detailed information could be displayed on landing and preview pages.

We also recommend that **SERP layouts** be extended beyond the traditional 10-blue- link paradigm to support more exploratory data search. Such interfaces might allow searching for and displaying key variables rather than entire datasets, as is already done within some data repositories. Another possibility would be to highlight links and similarities between different data sources during search and support comparisons during the search activity as is commonly done in product search. This further includes thinking about how to present data in search activities using voice assistants in conversational search settings. Implementing evaluation activities outside of visual or text interfaces might require thinking about different information needs for navigating data in this way.

6.1.3 Landing and Preview Pages

Landing and preview pages should include information based on particular data needs or data tasks, while also supporting deeper evaluation of data's relevance, usability and quality. To do this, landing/preview pages could contain **data summaries,** which could be textual, visual or statistical, as is already implemented in some portals. Such summaries could include data previews, column level summaries (in the case of structured data), or other ways to aggregate data which provide a general overview.

Summaries can help a user to understand the overall shape of a dataset in its entirety.

Depending on data needs, such summaries could also be used to emphasize certain aspects of the data. Summaries, drawing on relevant contextual information, could also be used to describe the provenance of an entire dataset (i.e. how data were collected or derived) or that of a particular part of the data.

Landing pages could also provide different avenues for "seeing into" the data. **Data previews**, which provide a sample of the data, have already been implemented in some data portals. These previews allow users to see what data look like and to begin to make sense of them as they evaluate the suitability of data for their needs. For spreadsheet data with multiple columns, previews could further include horizontal scrolling. Visualizations, as proposed by Marchionini et al. (2005) can also be used to support sensemaking.

As on SERPs, landing and preview pages should also provide ways to **zoom in and out** in order to facilitate different perspectives about the data, at different levels of detail. Interactive maps of geographic data or of entire collections could be one way to provide different representations of the data at different levels of granularity.

As mentioned, granular details about **location and time** are key in evaluating and understanding data for a particular task. While both geographic and temporal fields exist in common metadata schemas, they are not often captured or displayed in an intuitive way on landing or preview pages. Being able to zoom in to the desired level of detail (i.e. a street vs. a city vs. a country on a map) directly determines what a user can do with the data.

Information about the **history of the data,** including the data's original purpose, details of how they were created, and versioning information, should be present on landing and preview pages. Such data histories also include information about the choices and steps made during data cleaning, pre-processing and data sharing in order to enable reuse. There are also different ways of expressing column-based provenance, such as where the data in a column comes from, how it was created or from where it was derived.

Different **quality and uncertainty indicators** can be shown at this stage of the data discovery process as well. While data quality is dynamic and task-dependent, there are some shared standard metrics which could be displayed. Such indicators may include, but are not limited to, information about completeness or the presence of empty fields and headers. Creating a dashboard with a range of possible quality dimensions for users to select from could be one way to reflect the dynamic and personal nature of quality.

Landing and preview pages should also support evaluating data for **different types of tasks**, i.e. comparing or combining data. The data task and data use typologies presented in Chap. 3, e.g. the taxonomy proposed by Koesten et al. (2017) offer a starting point for thinking about different functionalities to offer. The five categories in this taxonomy include *linking, analyzing, summarizing, presenting* and *exporting*. Linking tasks, e.g., require comparing different data simultaneously; to support this, data should be viewed side-by-side and common attributes could be highlighted. Presenting and summarizing tasks could also be supported in various ways; we see this as a promising direction for future work in the space of human data interaction.

Data discovery system might also pre-empt analysis based on the task context. These could flag if the data are not suited for a task type, or if there is risk, such as issues of **representativeness**, or uncertainty, to be expected.

Data discovery is **linked with other search practices**, i.e. literature search. To support this, data discovery systems could be reciprocally linked to literature databases and data management tools. Landing pages should include literature citations with links for work which uses data; literature search systems should include data citations and links to data landing pages. This would not only provide context about how data have been used in the literature but would also enable data discovery from the literature.

Not all data uses will be cited in the academic literature. Links to other **instances where data have been used**, i.e. as a model input or in a data science workflow, could also be included on landing pages. **Glossaries** or codebooks which explain the terminology used within data, i.e. any abbreviations or definitions of variable names, can also support understanding.

In summary, we need to think about how to **contextualize data within data discovery**, by considering information both *from* and *about* data (as discussed in Chap. 5). As proposed before, a search index could be enriched with different types of context to allow algorithms to better retrieve datasets. We need to move away from the idea that data are isolated entities which can be discovered and understood without context.

6.1.4   Measure and Iterate

Thought needs to be given about how (and when) to set up the infrastructure and tools needed to capture and analyze user engagement and interaction flows within data discovery systems. Desired data indicators and metrics need to be discussed, defined and implemented in ways which enable insight over time. While potentially very useful, such metrics and indicators of use should not be considered out of context, and should be viewed as supplemental to qualitative insights from users themselves.

6.2   Recommendations for Metadata and Documentation

We have seen throughout this lecture that metadata are critical to both discovering and understanding data. We are by far not the first to emphasize the need for improved, standardized metadata to support data discovery (see Wilkinson et al. 2016). Here, we add our voices to these calls, but we also provide a different perspective by highlighting that metadata creation is itself a sociotechnical process. involving the work of both humans and machines.

We present recommendations and considerations for describing and documenting data in ways which support data discovery, sensemaking and interaction according to three categories (i) interoperability and open classifications, (ii) rich and linked documentation, and (iii) creation and curation.

6.2.1   Interoperability and Open Classifications

Echoing Wilkinson et al. we emphasize that data need to be described using common vocabularies and metadata standards which enable **machine readability as well as human understanding** (2016).

As shown in Chap. 3, data needs are diverse. Different metadata schemas can support different types of needs and activities. A general metadata standard may be enough to support finding data, but it is often not detailed enough to facilitate sensemaking. It is also perhaps unrealistic to believe that one metadata standard can fit all purposes and needs. This should be considered as repositories and data creators consider which schemas to use; **multiple types of metadata** may be needed to describe the same data.

Metadata standards need to be able to "speak" with each other, or to be interoperable. This can be achieved via, e.g., schema mapping or through methods incorporating Linked Open Data. Varying levels of granularity in metadata schemas and different definitions of fields and relationships can make such efforts challenging.

As work increases to make metadata standards interoperable, we need to remember that classifications are reductionist in their very nature. They obscure the messiness of the world, the data which they describe, and the decisions which are made during metadata creation (Bowker & Star, 1999). While there is no easy solution to this problem, one way forward could be to make some of this messiness visible.

To address this, the classifications which are incorporated into metadata schemas, i.e. for topic categorization or geographic locations, should be **well-documented and openly available**. This should be done in a way which supports human exploration and decisions about the meaning of classifications and how they are assigned in order to aid evaluation and sensemaking.

6.2.2 Rich and Linked Documentation

Data should be described using both standardized metadata as well as other forms of documentation. Links to resources which provide contextual information should also be provided to facilitate both discovery and sensemaking.

Metadata standards should include information which enable machines and humans to locate data in sustainable ways. **Persistent identifiers,** i.e. digital object identifiers (DOIs) should be assigned to data. **Data citations** should also include persistent identifiers and follow standard formats (as described in Groth et al. 2020). Citations to data provide both a means for people to locate data as well as a way to contextualize data reuse.

Other rich forms of documentation and linked contextual information (i.e. codebooks, ReadMe files, data summaries, and data histories with provenance information) should be provided alongside the data or be linked to the data.

Both metadata and other forms of documentation should allow people to evaluate the relevance, usability and quality of data, as discussed in Sect. 6.1. This may require **expanding the common minimum required metadata** fields to include information about topic, and when applicable, geographic and temporal information.

Potential privacy and ethical dimensions of data should be considered as documentation is created. Information about the types of uses to which data can be put, including restrictions for reuse put in place by either data creators or people whom the data describe, should be documented.

It can be difficult for descriptive information to be aggregated and displayed in new ways. However, metadata and documentation should be provided in ways which enable the computation and display of descriptive statistics or graphical summaries, as described in Sect. 6.1.

New ways of **co-creating data documentation** where both data creators and people using the data can contribute information should be explored. User-generated reviews and annotations could also be provided alongside data. Similarly, metadata schemas themselves should continue to evolve based on input from data repository managers, data creators, and the needs of people seeking and evaluating data for reuse.

6.2.3   Creation and Curation

Metadata and data documentation can be created automatically and manually. Certain metadata may be best captured using automated approaches. Technical advances in metadata enrichment and annotation can also be used to provide added context for fields such as subject, location and time.

Other information is best provided by humans. Data creators are often responsible for providing descriptions about data. Although they have the necessary knowledge about the data, they may not have the needed skills for creating metadata. **Trainings for metadata creation**, including motivating the practice as a way to enable data discoverability and reuse, offer one way forward.

Data creators also have unique insight about how data may efficiently be handled and understood. Data creators could **identify core, or anchor, variables** to aid others in navigating their data.

We have seen in Chap. 4 that data curation aids both discovery and reuse. This requires continued investment in curatorial efforts, as well as implementing automatic checks for metadata completeness and quality.

Curatorial and data management work is often hidden labor, which is not always recognized and valued. This is work performed not just by data managers or librarians, but is to some extent a part of all types of data work. Ways for recognizing and rewarding data management and documentation should continue to be investigated, as this work is what underlies discovering and making sense of data.

6.3   Recommendations for Collaborative Work and Community Engagement

We have seen throughout this lecture that data tasks are increasingly collaborative and that data discovery and sensemaking do not occur in isolation. Data seekers often do not find, access, and understand data purely by themselves.

Some of our earlier recommendations already build on the idea of supporting social interactions and on the importance of collaborative work. An open and difficult question is to consider how data discovery systems can further support collaborative work and community engagement.

Despite this, some recommendations can be made. Ways to contact data authors or to ask questions next to the data, i.e. in a forum or using a chat feature, could be implemented. Lists of clear tasks or questions could also be provided with the data. Questions could address information about the data, but also example questions that could be answered using the data. These questions and tasks could then be open to a community who could engage around them.

Discussion functionalities, tailored to data snippets, or screenshots of data, offer another way to engage communities and support collaboration. We can also consider how to better integrate existing knowledge about collaborative search and conversational search in communal settings.

Community rankings of data could also be a social signal in data discovery. Such rankings should also include contextual information about why particular data were up-voted or down-voted. Data seekers could also provide their own recommendations for similar data. We could also consider how to integrate data discovery into tools which already support collaborative work, or into the workflows of other data science tools.

Building on ideas from computer-supported scientific collaboration environments (Bos et al. 2007) there are many lessons that apply to specifically facilitating data-centric communities. Some are realized in repositories or online data science environments, such as Kaggle. Analyzing discussions around data online within such environments could help to further learn about how data are used and to identify common tasks and question types for data.

6.4 Recommendations for Data Seekers

This lecture focuses on presenting and understanding data discovery practices. Many of the recommendations we have made thus far are geared toward interface developers, systems designers, and data managers and curators. This work also has implications for people who are themselves looking for data.

General web search tips (Russell 2019), such as searching by file type, site, or image could also be used to locate data on the web. Tips more specific to searching for research data have also been proposed (Gregory et al. 2018). These include carefully considering how to construct queries to match specific data needs; selecting appropriate sources and making use of their bespoke functionalities; and considering looking for data services, where data are dynamically generated.

Adding to this, we urge data seekers to be aware of the limitations of online data search. When searching, they should be aware of the limitations in metadata and links, and to know that not all data may be findable using accustomed tools. We also recommend looking for data in a variety of sources, and to be aware that data discovery may take time and extensive research. Data seekers should also be aware that quality evaluations are personal, and that working with other people to discover and understand data may offer extra affordances that an online search alone does not.

Bibliography

Adams, A., & Blandford, A. (2005). Digital libraries' support for the user's "information journey." *Proceedings of the 5th ACM/IEEE-CS Joint Conference on Digital Libraries - JCDL '05*, 160. https://doi.org/10.1145/1065385.1065424

Agarwal, N. K. (2015). Towards a definition of serendipity in information behaviour. *Information Research.* https://doi.org/10.1016/i.bandc.2008.07.009

Agarwal, N. K. (2017). *Exploring context in information behavior: Seeker, situation, surroundings, and shared identities.* Morgan & Claypool Publishers. https://doi.org/10.2200/S00807ED1V01Y201710ICR061

Alrashed, T., Paparas, D., Benjelloun, O., Sheng, Y., & Noy, N. (2021). Dataset or not? A study on the veracity of semantic markup for dataset pages. In A. Hotho, E. Blomqvist, S. Dietze, A. Fokoue, Y. Ding, P. Barnaghi, A. Haller, M. Dragoni, & H. Alani (Eds.), *The Semantic Web – ISWC 2021* (pp. 338–356). Springer International Publishing. https://doi.org/10.1007/978-3-030-88361-4_20

Alter, G., Gonzalez-Beltran, A., Ohno-Machado, L., & Rocca-Serra, P. (2020). The Data Tags Suite (DATS) model for discovering data access and use requirements. *GigaScience, 9*(2), giz165. https://doi.org/10.1093/gigascience/giz165

Ames, D. P., Horsburgh, J. S., Cao, Y., Kadlec, J., Whiteaker, T., & Valentine, D. (2012). HydroDesktop: Web services-based software for hydrologic data discovery, download, visualization, and analysis. *Environmental Modelling and Software, 37*, 146–156. https://doi.org/10.1016/j.envsoft.2012.03.013

Araujo, P. C. de, Castanha, R. C. G., & Hjorland, B. (2021). Citation indexing and indexes. *Knowledge Organization, 48*(1), 72–101.

Ariño, A. H., Chavan, V., & Faith, D. P. (2013). Assessment of user needs of primary biodiversity data: Analysis, concerns, and challenges. *Biodiversity Informatics, 8*(2), Article 2. https://doi.org/10.17161/bi.v8i2.4094

Atz, U. (2014). The tau of data: A new metric to assess the timeliness of data in catalogues. *Conference for E-Democracy and Open Governement*, 257.

Au, V., Thomas, P., & Jayasinghe, G. K. (2016). Query-biased summaries for tabular data. *Proceedings of the 21st Australasian Document Computing Symposium*, 69–72. https://doi.org/10.1145/3015022.3015027

Baker, J., Jones, D. R., & Burkman, J. (2009). Using visual representations of data to enhance sensemaking in data exploration task. *Journal of the Association for Information Systems, 10*(7).

Baker, K. S., Duerr, R. E., & Parsons, M. A. (2015). Scientific knowledge mobilization: Co-evolution of data products and designated communities. *International Journal of Digital Curation, 10*(2), 110–135. https://doi.org/10.2218/ijdc.v10i2.346

Bales, S., & Wang, P. (2005). Consolidating user relevance criteria: A meta-ethnography of empirical studies. *Proceedings of the American Society for Information Science and Technology, 42*(1). https://doi.org/10.1002/meet.14504201277

Balog, K., Meij, E., & de Rijke, M. (2010). Entity search: Building bridges between two worlds. *Proceedings of the 3rd International Semantic Search Workshop,* 9:1–9:5. https://doi.org/10.1145/1863879.1863888

Bando, L. L., Scholer, F., & Turpin, A. (2010). Constructing query-biased summaries: A comparison of human and system generated snippets. *Proceedings of the Third Symposium on Information Interaction in Context,* 195–204. https://doi.org/10.1145/1840784.1840813

Barry, C. L. (1994). User-defined relevance criteria: An exploratory study. *JASIS, 45*(3), 149–159. https://doi.org/10.1002/(SICI)1097-4571(199404)45:3<149::AID-ASI5>3.0.CO;2-J

Bates, M. J. (1989). The design of browsing and berrypicking techniques for the online search interface. *Online Review, 13*(5), 407–424.

Batini, C., Cappiello, C., Francalanci, C., & Maurino, A. (2009). Methodologies for data quality assessment and improvement. *ACM Computing Surveys, 41*(3), 16:1–16:52. https://doi.org/10.1145/1541880.1541883

Bautista-Puig, N., De Filippo, D., Mauleón, E., & Sanz-Casado, E. (2019). Scientific landscape of citizen science publications: Dynamics, content and presence in social media. *Publications, 7*(1), 12. https://doi.org/10.3390/publications7010012

Belkin, N. J. (1993). Interaction with texts: Information retrieval as information-seeking behavior. *Information Retrieval,* 55–66.

Belkin, N. J. (1996). Intelligent information retrieval: Whose intelligence? *ISI '96: Proceedings of the Fifth International Symposium for Information Science,* 25–31.

Ben Ellefi, M., Bellahsene, Z., Dietze, S., & Todorov, K. (2016). Dataset recommendation for data linking: An intensional approach. In H. Sack, E. Blomqvist, M. d'Aquin, C. Ghidini, S. P. Ponzetto, & C. Lange (Eds.), *The Semantic Web. Latest Advances and New Domains* (Vol. 9678, pp. 36–51). Springer International Publishing. https://doi.org/10.1007/978-3-319-34129-3_3

Benjelloun, O., Chen, S., & Noy, N. (2020). Google dataset search by the numbers. In J. Z. Pan, V. Tamma, C. d'Amato, K. Janowicz, B. Fu, A. Polleres, O. Seneviratne, & L. Kagal (Eds.), *The Semantic Web – ISWC 2020* (pp. 667–682). Springer International Publishing. https://doi.org/10.1007/978-3-030-62466-8_41

Beran, B., Cox, S. J. D., Valentine, D., Zaslavsky, I., & McGee, J. (2009). Web services solutions for hydrologic data access and cross-domain interoperability. *International Journal on Advances in Intelligent Systems, 2*(2 & 3).

Birnholtz, J. P., & Bietz, M. J. (2003). Data at work: Supporting sharing in science and engineering. In K. Schmidt, M. Pendergast, M. Tremaine, & C. Simone (Eds.), *Proceedings of the 2003 International ACM SIGGROUP Conference on Supporting Group Work, GROUP 2003, Sanibel Island, Florida, USA, November 9–12, 2003* (pp. 339–348). ACM. https://doi.org/10.1145/958160.958215

Bishop, B. W., & Hank, C. (2018). Measuring FAIR Principles to inform fitness for use. *IJDC, 13*(1), 35–46. https://doi.org/10.2218/ijdc.v13i1.630

Bishop, B. W., Hank, C., Webster, J., & Howard, R. (2019). Scientists' data discovery and reuse behavior: (Meta)data fitness for use and the FAIR data principles. *Proceedings of the Association for Information Science and Technology, 56*(1), 21–31. https://doi.org/10.1002/pra2.4

Bishop, L. (2009). Ethical sharing and reuse of qualitative data. *Australian Journal of Social Issues, 44*(3). https://doi.org/10.1002/j.1839-4655.2009.tb00145.x

Bishop, L., & Kuula-Luumi, A. (2017). Revisiting qualitative data reuse: A decade on. *SAGE Open*, *7*(1), 2158244016685136. https://doi.org/10.1177/2158244016685136

Blandford, A., & Attfield, S. (2010). *Interacting with Information*. Morgan & Claypool Publishers. https://doi.org/10.2200/S00227ED1V01Y200911HCI006

Borgman, C. L. (2012). The conundrum of sharing research data. *Journal of the American Society for Information Science and Technology*, *63*(6), 1059–1078. https://doi.org/10.1002/asi.22634

Borgman, C. L. (2015). *Big data, little data, no data: Scholarship in the networked world*. MIT press. http://www.jstor.org/stable/j.ctt17kk8n8

Borgman, C. L., Scharnhorst, A., & Golshan, M. S. (2019). Digital data archives as knowledge infrastructures: Mediating data sharing and reuse. *Journal of the Association for Information Science and Technology*, *70*(8), 888–904. https://doi.org/10.1002/asi.24172

Borgman, C. L., Smart, L. J., Millwood, K. A., Finley, J. R., Champeny, L., Gilliland, A. J., & Leazer, G. H. (2005). Comparing faculty information seeking in teaching and research: Implications for the design of digital libraries. *Journal of the American Society for Information Science and Technology*, *56*(6), 636–657. https://doi.org/10.1002/asi.20154

Borgman, C. L., Van de Sompel, H., Scharnhorst, A., van den Berg, H., & Treloar, A. (2015). Who uses the digital data archive? An exploratory study of DANS. *Proceedings of the Association for Information Science and Technology*, *52*, 1–5. https://doi.org/10.1002/pra2.2015.145052010096

Borgman, C. L., Wallis, J. C., Mayernik, M. S., & Pepe, A. (2007). Drowning in data: Digital library architecture to support scientific use of embedded sensor networks. *Proceedings of the 7th ACM/IEEE-CS Joint Conference on Digital Libraries*, 269–277. https://doi.org/10.1145/1255175.1255228

Borgman, C. L., Wofford, M. F., Golshan, M. S., & Darch, P. T. (2021). Collaborative qualitative research at scale: Reflections on 20 years of acquiring global data and making data global. *Journal of the Association for Information Science and Technology*, *72*(6), 667–682. https://doi.org/10.1002/asi.24439

Bos, N., Zimmerman, A., Olson, J., Yew, J., Yerkie, J., Dahl, E., & Olson, G. (2007). From shared databases to communities of practice: A Taxonomy of Collaboratories. *Journal of Computer-Mediated Communication*, *12*(2), 652–672. https://doi.org/10.1111/j.1083-6101.2007.00343.x

Boukhelifa, N., Perrin, M.-E., Huron, S., & Eagan, J. (2017). How data workers cope with uncertainty: A task characterisation study. *Proceedings of the 2017 CHI Conference on Human Factors in Computing Systems*, 3645–3656. https://doi.org/10.1145/3025453.3025738

Bowker, G. C., & Star, S. L. (1999). *Sorting things out: Classification and its consequences*. The MIT Press. ISBN: 9780262522953

Brickley, D., Burgess, M., & Noy, N. F. (2019). Google Dataset Search: Building a search engine for datasets in an open Web ecosystem. *The World Wide Web Conference, WWW 2019, San Francisco, CA, USA, May 13–17, 2019*, 1365–1375. https://doi.org/10.1145/3308558.3313685

Bron, M., Balog, K., & de Rijke, M. (2010). Ranking related entities: Components and analyses. *Proceedings of the 19th ACM International Conference on Information and Knowledge Management*, 1079–1088. https://doi.org/10.1145/1871437.1871574

Brown, C. (2003). The changing face of scientific discourse: Analysis of genomic and proteomic database usage and acceptance. *Journal of the American Society for Information Science and Technology*, *54*(10), 926–938. https://doi.org/10.1002/asi.10289

Buckland, M. K. (1991). Information as thing. *Journal of the American Society for Information Science*, *42*(5), 351–360. https://doi.org/10.1002/(SICI)1097-4571(199106)42:5<351::AID-ASI5>3.0.CO;2-3

Burton-Taylor. (2015). Demand for financial market data and news. In *Report*. Burton-Taylor.

Cafarella, M. J., Halevy, A., & Madhavan, J. (2011). Structured data on the Web. *Commun. ACM*, *54*(2), 72–79. https://doi.org/10.1145/1897816.1897839

Cafarella, M. J., Halevy, A. Y., Wang, D. Z., Wu, E., & Zhang, Y. (2008). WebTables: Exploring the power of tables on the web. *PVLDB, 1*(1), 538–549.

Case, D. O., & Given, L. M. (2016). *Looking for information: A survey of research on information seeking, needs, and behavior* (Fourth edition). Emerald Group Publishing Limited.

Cattaneo, G., Glennon, M., Lifonti, R., Giorgio Micheletti, Woodward, A., Kolding, M., Vacca, A., Croce, C. L., & Osimo, D. (2015). *European Data Market SMART 2013/0063, D6—First Interim Report.* https://idc-emea.app.box.com/s/k7xv0u3gl6xfvq1rl667xqmw69pzk790

Chapman, A., Simperl, E., Koesten, L., Konstantinidis, G., Ibáñez, L.-D., Kacprzak, E., & Groth, P. (2020). Dataset search: A survey. *The VLDB Journal, 29*(1), 251–272. https://doi.org/10.1007/s00778-019-00564-x

Clarivate. (2022). Data Citation Index. *Data Citation Index.* https://clarivate.com/webofscience group/solutions/webofscience-data-citation-index

Convertino, G., & Echenique, A. (2017). Self-service data preparation and analysis by business users: New needs, skills, and tools. *Proceedings of the 2017 CHI Conference Extended Abstracts on Human Factors in Computing Systems*, 1075–1083. https://doi.org/10.1145/3027063.3053359

Courtright, C. (2007). Context in information behavior research. *Annual Review of Information Science and Technology, 41*(1), 273–306. https://doi.org/10.1002/aris.2007.1440410113

Curty, R. G. (2016). Factors influencing research data reuse in the social sciences: An exploratory study. *International Journal of Digital Curation, 11*(1), 96–117. https://doi.org/10.2218/ijdc.v11 i1.401

Dai, Z., Kim, Y., & Callan, J. (2017). Learning to rank resources. *Proceedings of the 40th International ACM SIGIR Conference on Research and Development in Information Retrieval*, 837–840. https://doi.org/10.1145/3077136.3080657

Degbelo, A. (2020). Open Data user needs: A preliminary synthesis. *Companion Proceedings of the Web Conference 2020*, 834–839. https://doi.org/10.1145/3366424.3386586

Dervin, B. (1997). Given a context by any other name: Methodological tools for taming the unruly beast. In P. Vakkari, R. Savolainen, & B. Dervin (Eds.), *Information Seeking in Context* (pp. 13–38). Taylor Graham.

Dervin, B. (1983). *An overview of sense-making: Concepts, methods, and results to date.* Annual Meeting of International Communication Association, Dallas. https://www.ideals.illinois.edu/bit stream/handle/2142/2281/Dervin83a.htm

Dervin, B. (1998). Sense-making theory and practice: An overview of user interests in knowledge seeking and use. *Journal of Knowledge Management, 2*(2), 36–46. https://doi.org/10.1108/136 73279810249369

Dervin, B., & Nilan, M. (1986). Information needs and uses. *Annual Review of Information Science and Technology, 21*, 3–33.

Deutsches Institut für Normung. (2012). *DIN31644:2012–04.* Information and Documentation - Criteria for Trustworthy Digital Archives. https://www.beuth.de/en/standard/din-31644/147 058907

Dillo, I., & Leeuw, L. de. (2018). CoreTrustSeal. *Mitteilungen Der Vereinigung Österreichischer Bibliothekarinnen & Bibliothekare, 71*(1), 162–170. http://eprints.rclis.org/34431/

Directorate-General for Research and Innovation (European Commission). (2021). *Horizon Europe, open science: Early knowledge and data sharing, and open collaboration.* Publications Office of the European Union. https://doi.org/10.2777/18252

Dixit, R., Rogith, D., Narayana, V., Salimi, M., Gururaj, A., Ohno-Machado, L., Xu, H., & Johnson, T. R. (2018). User needs analysis and usability assessment of DataMed – a biomedical data discovery index. *Journal of the American Medical Informatics Association, 25*(3), 337–344. https://doi.org/10.1093/jamia/ocx134

Dourish, P. (2004). What we talk about when we talk about context. *Personal and Ubiquitous Computing, 8*(1), 19–30. https://doi.org/10.1007/s00779-003-0253-8

Dourish, P., & Gómez Cruz, E. (2018). Datafication and data fiction: Narrating data and narrating with data. *Big Data & Society, 5*(2), 2053951718784083. https://doi.org/10.1177/2053951718784083

Dow, A. K., Dow, E. M., Fitzsimmons, T. D., & Materise, M. M. (2015). Harnessing the environmental data flood: A comparative analysis of hydrologic, oceanographic, and meteorological informatics platforms. *Bulletin of the American Meteorological Society, 96*(5), 725–736. https://doi.org/10.1175/BAMS-D-13-00178.1

Edmond, J., & Nugent Folan, G. (2017). Data, metadata, narrative. Barriers to the reuse of cultural sources. In E. Garoufallou, S. Virkus, R. Siatri, & D. Koutsomiha (Eds.), *Metadata and Semantic Research* (pp. 253–260). Springer International Publishing. https://doi.org/10.1007/978-3-319-70863-8_25

Edwards, P. N., Mayernik, M. S., Batcheller, A. L., Bowker, G. C., & Borgman, C. L. (2011). Science friction: Data, metadata, and collaboration. *Social Studies of Science, 41*(5), 667–690. https://doi.org/10.1177/0306312711413314

Ellis, D. (1989). A behavioural approach to information retrieval system design. *Journal of Documentation, 45*(3), 171–212. https://doi.org/10.1108/eb026843

Ellis, D., Cox, D., & Hall, K. (1993). A comparison of the information seeking patterns of researchers in the physical and social sciences. *Journal of Documentation, 49*(4), 356–369. https://doi.org/10.1108/eb026919

Ellis, D., & Haugan, M. (1997). Modelling the information seeking patterns of engineers and research scientists in an industrial environment. *Journal of Documentation, 53*(4), 384–403. https://doi.org/10.1108/EUM0000000007204

European Commission. (2011). *Digital Agenda: Commission's Open Data Strategy, Questions & Answers* [Text]. https://ec.europa.eu/commission/presscorner/detail/en/MEMO_11_891

European Commission. (2018). *Turning FAIR into reality. Final report and action plan from the European Commission expert group on FAIR data.* European Commission. Directorate General for Research and Innovation. Directorate B Open Innovation and Open Science. Unit B2 Open Science. https://ec.europa.eu/info/sites/info/files/turning_fair_into_reality_1.pdf

Faniel, I. M., Austin, A., Kansa, E., Kansa, S. W., France, P., Jacobs, J., Boytner, R., & Yakel, E. (2018). Beyond the archive bridging data creation and reuse in archaeology. *Advances in Archaeological Practice, 6*(2), 105–116. https://doi.org/10.1017/aap.2018.2

Faniel, I. M., Barrera-Gomez, J., Kriesberg, A., & Yakel, E. (2013a). A comparative study of data reuse among quantitative social scientists and archaeologists. *IConference 2013a Proceedings,* 797–800. https://doi.org/10.9776/13391

Faniel, I. M., Frank, R. D., & Yakel, E. (2019). Context from the data reuser's point of view. *Journal of Documentation, 75*(6), 1274–1297. https://doi.org/10.1108/JD-08-2018-0133

Faniel, I. M., Kansa, E., Kansa, S. W., Barrera-Gomez, J., & Yakel, E. (2013b). The challenges of digging data: A study of context in archaeological data reuse. *13th ACM/IEEE-CS Joint Conference on Digital Libraries, JCDL '13, Indianapolis, IN, USA, July 22 - 26, 2013b,* 295–304. https://doi.org/10.1145/2467696.2467712

Faniel, I. M., & Yakel, E. (2017). Practices do not make perfect: Disciplinary data sharing and reuse practices and their implications for repository data curation. In *Curating research data, volume one: Practical strategies for your digital repository* (pp. 103–126). Association of College and Research Libraries Chicago, Illinois.

Fear, K. (2013). *Measuring and anticipating the impact of data reuse* [University of Michigan]. https://deepblue.lib.umich.edu/handle/2027.42/102481

Federer, L. M. (2019). Who, what, when, where, and why? Quantifying and understanding biomedical data reuse. https://drum.lib.umd.edu/handle/1903/21991

Floridi, L. (2010). *Information: A very short introduction* (Issue Vol. 225). Oxford University Press. https://doi.org/10.1093/actrade/9780199551378.001.0001

Force, M. M., & Robinson, N. J. (2014). Encouraging data citation and discovery with the Data Citation Index. *Journal of Computer-Aided Molecular Design, 28*(10), 1043–1048. https://doi.org/10.1007/s10822-014-9768-5

Freitas, A., O'Riáin, S., & Curry, E. (2020). Querying and Searching Heterogeneous Knowledge Graphs in Real-time Linked Dataspaces. In E. Curry, *Real-time Linked Dataspaces* (pp. 105–124). Springer International Publishing. https://doi.org/10.1007/978-3-030-29665-0_7

Freund, L. (2013). A cross-domain analysis of task and genre effects on perceptions of usefulness. *Inf. Process. Manage., 49*(5), 1108–1121. https://doi.org/10.1016/j.ipm.2012.08.007

Friedrich, T. (2020). *Looking for data: Information seeking behaviour of survey data users.* [Humboldt University]. https://edoc.hu-berlin.de/handle/18452/22813

Furnas, G. W., & Russell, D. M. (2005). Making sense of sensemaking. *Extended Abstracts Proceedings of the 2005 Conference on Human Factors in Computing Systems, CHI 2005, Portland, Oregon, USA, April 2–7, 2005*, 2115–2116. https://doi.org/10.1145/1056808.1057113

Garnter. (2022). *Data and analytics: Everything you need to know.* https://www.gartner.com/en/topics/data-and-analytics

Gebru, T., Morgenstern, J., Vecchione, B., Vaughan, J. W., Wallach, H., III, H. D., & Crawford, K. (2021). Datasheets for datasets. *Communications of the ACM, 64*(12), 86–92. https://doi.org/10.1145/3458723

Ghasemaghaei, M., & Calic, G. (2019). Does big data enhance firm innovation competency? The mediating role of data-driven insights. *Journal of Business Research, 104*, 69–84. https://doi.org/10.1016/j.jbusres.2019.07.006

Gilliland, A. (2008). Setting the stage. In M. Baca & Getty Research Institute (Eds.), *Introduction to metadata* (pp. 1–19). Getty Research Institute. https://doi.org/10.1177/155019061701300104

Gonzalez, H., Halevy, A. Y., Jensen, C. S., Langen, A., Madhavan, J., Shapley, R., Shen, W., & Goldberg-Kidon, J. (2010). Google fusion tables: Web-centered data management and collaboration. *Proceedings of the 2010 ACM SIGMOD International Conference on Management of Data*, 1061–1066. https://doi.org/10.1145/1807167.1807286

Goyal, N., & Fussell, S. R. (2016). Effects of sensemaking translucence on distributed collaborative analysis. *Proceedings of the 19th ACM Conference on Computer-Supported Cooperative Work & Social Computing*, 288–302. https://doi.org/10.1145/2818048.2820071

Gregory, K. (2021). *Findable and reusable? Data discovery practices in research* [Maastricht University]. https://doi.org/10.26481/dis.20210302kg

Gregory, K., Groth, P., Cousijn, H., Scharnhorst, A., & Wyatt, S. (2019). Searching data: A review of observational data retrieval practices in selected disciplines. *Journal of the Association for Information Science and Technology, 70*(5), 419–432. https://doi.org/10.1002/asi.24165

Gregory, K., Groth, P., Scharnhorst, A., & Wyatt, S. (2020a). Lost or Found? Discovering Data Needed for Research. *Harvard Data Science Review.* https://doi.org/10.1162/99608f92.e38165eb

Gregory, K., Khalsa, S. J., Michener, W. K., Psomopoulos, F. E., de Waard, A., & Wu, M. (2018). Eleven quick tips for finding research data. *PLOS Computational Biology, 14*(4), e1006038. https://doi.org/10.1371/journal.pcbi.1006038

Gregory, K. M., Cousijn, H., Groth, P., Scharnhorst, A., & Wyatt, S. (2020b). Understanding data search as a socio-technical practice. *Journal of Information Science, 0*(0), 0165551519837182. https://doi.org/10.1177/0165551519837182

Groth, P., Cousijn, H., Clark, T., & Goble, C. (2020). FAIR data reuse–the path through data citation. *Data Intelligence*, 78–86. https://doi.org/10.1162/dint_a_00030

Koesten, L., Mayr, P., Groth, P., Simperl, E., & de Rijke, M. (2019). Report on the DATA: SEARCH'18 workshop - Searching Data on the Web. *ACM SIGIR Forum*, *52*(2), 117–124. https://doi.org/10.1145/3308774.3308794

Halevy, A. Y., Korn, F., Noy, N. F., Olston, C., Polyzotis, N., Roy, S., & Whang, S. E. (2016). Goods: Organizing Google's datasets. *Proceedings of the 2016 International Conference on Management of Data, SIGMOD Conference 2016, San Francisco, CA, USA, June 26 - July 01, 2016*, 795–806. https://doi.org/10.1145/2882903.2903730

Hartig, O., Bizer, C., & Freytag, J.-C. (2009). Executing SPARQL queries over the web of Linked Data. In A. Bernstein, D. R. Karger, T. Heath, L. Feigenbaum, D. Maynard, E. Motta, & K. Thirunarayan (Eds.), *The Semantic Web—ISWC 2009* (Vol. 5823, pp. 293–309). Springer Berlin Heidelberg. https://doi.org/10.1007/978-3-642-04930-9_19

Hearst, M. (2009). *Search user interfaces*. Cambridge University Press. ISBN 9780521113793

Heimgärtner, R. (2020, July). Development of an assessment model for the human centered design processes specified in ISO 9241-220. In *International Conference on Human-Computer Interaction* (pp. 50–70). Springer, Cham.

Hendler, J., Holm, J., Musialek, C., & Thomas, G. (2012). US Government Linked Open Data: Semantic.data.gov. *IEEE Intelligent Systems*, *27*(3), 25–31. https://doi.org/10.1109/MIS.2012.27

Hermans, F., Pinzger, M., & van Deursen, A. (2011). Supporting professional spreadsheet users by generating leveled dataflow diagrams. *Proceedings of the 33rd International Conference on Software Engineering*, 451–460. https://doi.org/10.1145/1985793.1985855

Hersh, W., & Voorhees, E. (2009). TREC genomics special issue overview. *Information Retrieval*, *12*(1), 1–15. https://doi.org/10.1007/s10791-008-9076-6

Hey, T., Tansley, S., & Tolle, K. M. (2009). Jim Gray on eScience: A transformed scientific method. In *The Fourth Paradigm: Data-Intensive Scientific Discovery*. http://research.microsoft.com/en-us/collaboration/fourthparadigm/4th_paradigm_book_jim_gray_transcript.pdf

Hienert, D., Kern, D., Boland, K., Zapilko, B., & Mutschke, P. (2019). A digital library for research data and related information in the social sciences. *2019 ACM/IEEE Joint Conference on Digital Libraries (JCDL)*, 148–157. https://doi.org/10.1109/JCDL.2019.00030

Hirschman, L., & Gaizauskas, R. (2001). Natural language question answering: The view from here. *Natural Language Engineering*, *7*(4), 275–300. https://doi.org/10.1017/S1351324901002807

Hogan, A., Harth, A., Umbrich, J., Kinsella, S., Polleres, A., & Decker, S. (2011). Searching and browsing Linked Data with SWSE: The Semantic Web Search Engine. *Journal of Web Semantics*, *9*(4), 365–401. https://doi.org/10.1016/j.websem.2011.06.004

Holland, S., Hosny, A., Newman, S., Joseph, J., & Chmielinski, K. (2018). The Dataset Nutrition Label: A framework to drive higher data quality standards. *CoRR*, *abs/1805.03677*. http://arxiv.org/abs/1805.03677

Ibáñez, L.-D., Koesten, L., Kacprzak, E., & Simperl, E. (2020). *Characterising Dataset Search on the European Data Portal | data.europa.eu* (Analytical Report 18; pp. 1–42). European Data Portal. https://data.europa.eu/en/datastories/analytical-report-18-characterising-dataset-search-european-data-portal

Ingwersen, P. (1996). Cognitive perspectives of information retrieval interaction: Elements of a cognitive IR theory. *Journal of Documentation*, *52*(1), 3–50.

Ingwersen, P. E. R. (1992). *Information Retrieval Interaction*. Taylor Graham. ISBN-13: 978-0947568542

Ingwersen, P., & Järvelin, K. (2005). *The turn: Integration of information seeking and retrieval in context*. Springer. ISBN: 978-1-4020-3851-8

Juran, J. M., & De Feo, J. A. (Eds.). (2017). *Juran's quality handbook: The complete guide to performance excellence* (Seventh edition). McGraw Hill Education. https://www.accessengineeringlibrary.com/content/book/9781259643613

Kacprzak, E., Koesten, L., Ibáñez, L.-D., Blount, T., Tennison, J., & Simperl, E. (2019). Characterising dataset search—An analysis of search logs and data requests. *Journal of Web Semantics*. https://doi.org/10.1016/j.websem.2018.11.003

Kacprzak, E., Koesten, L. M., Ibáñez, L.-D., Simperl, E., & Tennison, J. (2017). A query log analysis of dataset search. In J. Cabot, R. De Virgilio, & R. Torlone (Eds.), *Web Engineering. ICWE 2017.* (Vol. 10360, pp. 429–436). Springer, Cham. https://doi.org/10.1007/978-3-319-60131-1_29

Kammerer, Y., & Gerjets, P. (2010). How the interface design influences users' spontaneous trustworthiness evaluations of web search results: Comparing a list and a grid interface. *Proceedings of the 2010 Symposium on Eye-Tracking Research & Applications*, 299–306. https://doi.org/10.1145/1743666.1743736

Kang, Y. & Stasko, J. T. (2012). Examining the use of a visual analytics system for sensemaking tasks: Case studies with domain experts. *IEEE Transactions on Visualization and Computer Graphics*, *18*(12), 2869–2878

Karkouch, A., Mousannif, H., Al Moatassime, H., & Noel, T. (2016). Data quality in internet of things: A state-of-the-art survey. *Journal of Network and Computer Applications*, *73*, 57–81. https://doi.org/10.1016/j.jnca.2016.08.002

Kashfi, P., Nilsson, A., & Feldt, R. (2017). Integrating User eXperience practices into software development processes: Implications of the UX characteristics. *PeerJ Computer Science*, *3*, e130. https://doi.org/10.7717/peerj-cs.130

Kassen, M. (2013). A promising phenomenon of open data: A case study of the Chicago open data project. *Government Information Quarterly*, *30*(4), 508–513. https://doi.org/10.1016/j.giq.2013.05.012

Kato, M., Ohshima, H., Liu, Y.-H., & Chen, H. (2020). Overview of the NTCIR-15 Data Search Task. *Proceedings of the 15th NTCIR Conference on Evaluation of Information Access Technologies*. NTCIR 15 Conference, Tokyo, Japan. ISBN: 978-4-86049-080-5

Kelly, D. (2009). Methods for evaluating interactive information retrieval systems with users. *Foundations and Trends in Information Retrieval*, *3*(1–2), 1–224. https://doi.org/10.1561/1500000012

Kelly, D., & Azzopardi, L. (2015). How many results per page? A Study of SERP size, search behavior and user experience. *Proceedings of the 38th International ACM SIGIR Conference on Research and Development in Information Retrieval*, 183–192. https://doi.org/10.1145/2766462.2767732

Kennedy, H. (2018). Living with data: Aligning data studies and data activism through a focus on everyday experiences of datafication. *Krisis : Journal for Contemporary Philosophy*, *2018*(1), 18–30. ISSN 1875-7103 (Online)

Kern, D., & Mathiak, B. (2015). Are there Any differences in data set retrieval compared to well-known literature retrieval? *Research and Advanced Technology for Digital Libraries - 19th International Conference on Theory and Practice of Digital Libraries, TPDL 2015, Poznań, Poland, September 14–18, 2015. Proceedings*, 197–208. https://doi.org/10.1007/978-3-319-24592-8_15

Key Perspectives. (2010). *Data dimensions: Disciplinary differences in research data sharing, reuse and long term viability. SCARP Synthesis Study*. Digital Curation Centre. http://www.dcc.ac.uk/scarp

Khan, A., Tiropanis, T., & Martin, D. (2016). Exploiting semantic annotation of content with Linked Open Data (LoD) to improve searching performance in web repositories of multi-disciplinary research data. In Braslavski P. et al. (Ed.), *Information Retrieval. RuSSIR 2015. Communications in Computer and Information Science* (Vol. 573, pp. 130–145). https://doi.org/10.1007/978-3-319-41718-9_7

Kitchin, R., & McArdle, G. (2016). What makes Big Data, Big Data? Exploring the ontological characteristics of 26 datasets. *Big Data & Society*, *3*(1), 2053951716631130. https://doi.org/10.1177/2053951716631130

Klein, G. A., Orasanu, J., Calderwood, R., Zsambok, C. E., & others. (1993). *Decision making in action: Models and methods*. Ablex Norwood, NJ.

Klein, G., Moon, B., & Hoffman, R. R. (2006a). Making sense of sensemaking 2: A Macrocognitive Model. *IEEE Intelligent Systems*, *21*(5), 88–92. https://doi.org/10.1109/MIS.2006.100

Klein, G., Moon, B. M., & Hoffman, R. R. (2006b). Making sense of sensemaking 1: Alternative Perspectives. *IEEE Intelligent Systems*, *21*(4), 70–73. https://doi.org/10.1109/MIS.2006.75

Klein, G., Phillips, J. K., Rall, E. L., & Peluso, D. A. (2007). A data–frame theory of sensemaking. In *Expertise out of context* (pp. 118–160). Psychology Press. eBook ISBN: 9780429235481

Knorr Cetina, K. (1999). *Epistemic cultures: How the sciences make knowledge*. Harvard University Press. https://doi-org.uaccess.univie.ac.at/10.2307/j.ctvxw3q7f

Koesten, L. (2019). *A user centred perspective on structured data discovery*. University of Southampton, Faculty of Engineering, Science and Mathematics. https://eprints.soton.ac.uk/438583/1/Final_thesis.pdf

Koesten, L. M., Kacprzak, E., Tennison, J. F. A., & Simperl, E. (2017). The trials and tribulations of working with structured data: A study on information seeking behaviour. *Proceedings of the 2017 CHI Conference on Human Factors in Computing Systems*, 1277–1289. https://doi.org/10.1145/3025453.3025838

Koesten, L., Kacprzak, E., Tennison, J., & Simperl, E. (2019b). Collaborative practices with structured data: Do tools support what users need? *Proceedings of the 2019b CHI Conference on Human Factors in Computing Systems, CHI 2019b, Glasgow, Scotland, UK, May 04–09, 2019b*, 100. https://doi.org/10.1145/3290605.3300330

Koesten, L., & Simperl, E. (2021). UX of data: Making data available doesn't make it usable. *Interactions*, *28*(2), 97–99. https://doi.org/10.1145/3448888

Koesten, L., Simperl, E., Blount, T., Kacprzak, E., & Tennison, J. (2019c). Everything you always wanted to know about a dataset: Studies in data summarisation. *International Journal of Human-Computer Studies*. https://doi.org/10.1016/j.ijhcs.2019.10.004

Koesten, L., Vougiouklis, P., Simperl, E., & Groth, P. (2020). Dataset reuse: Toward translating principles to practice. *Patterns*, *1*(8), 100136. https://doi.org/10.1016/j.patter.2020.100136

Koesten, L., Gregory, K., Groth, P., & Simperl, E. (2021). Talking datasets – Understanding data sensemaking behaviours. *International Journal of Human-Computer Studies*, *146*, 102562. https://doi.org/10.1016/j.ijhcs.2020.102562

Krämer, T., Papenmeier, A., Carevic, Z., Kern, D., & Mathiak, B. (2021). Data-seeking behaviour in the social sciences. *International Journal on Digital Libraries*. https://doi.org/10.1007/s00799-021-00303-0

Kriesberg, A., Frank, R. D., Faniel, I. M., & Yakel, E. (2013). The role of data reuse in the apprenticeship process. *Proceedings of the American Society for Information Science and Technology*, *50*, 1–10. https://doi.org/10.1002/meet.14505001051

Kuhn, T. S. (1962). *The structure of scientific revolutions*. University of Chicago Press.

Lane, J., Mulvany, I., & Paco, N. (Eds.). (2020). *Rich search and discovery for research datasets*. Sage Publishing. https://study.sagepub.com/richcontext

Leonelli, S. (2015). What counts as scientific data? A relational framework. *Philosophy of Science*, *82*(5), 810–821. https://doi.org/10.1086/684083

Leonelli, S. (2016). *Data-centric biology: A philosophical study*. University of Chicago Press.

Leonelli, S., & Ankeny, R. A. (2015). Repertoires: How to transform a project into a research community. *BioScience*, *65*(7), 701–708. https://doi.org/10.1093/biosci/biv061

Lievrouw, L. A. (2001). New media and the pluralization of life-worlds': A role for information in social differentiation. *New Media & Society, 3*(1), 7–28. https://doi.org/10.1177/146144480100 3001002

Lillis, T. (2011). Legitimizing dialogue as textual and ideological goal in academic writing for assessment and publication. *Arts and Humanities in Higher Education, 10*(4), 401–432. https:// doi.org/10.1177/1474022211398106

Lillis, T., & Maitlis, S. (2005). The social processes of organizational sensemaking. *Arts and Humanities in Higher Education, 48*(1), 21–49. https://doi.org/10.1177/1474022211398106

Lin, D., Crabtree, J., Dillo, I., Downs, R. R., Edmunds, R., Giaretta, D., De Giusti, M., L'Hours, H., Hugo, W., Jenkyns, R., Khodiyar, V., Martone, M. E., Mokrane, M., Navale, V., Petters, J., Sierman, B., Sokolova, D. V., Stockhause, M., & Westbrook, J. (2020). The TRUST Principles for digital repositories. *Scientific Data, 7*(1), 144. https://doi.org/10.1038/s41597-020-0486-7

Liu, Y.-H., Chen, H.-L. (Oliver), Kato, M. P., Wu, M., & Gregory, K. (2021). Data discovery and reuse in data service practices: A global perspective. *Proceedings of the Association for Information Science and Technology, 58*(1), 610–612. https://doi.org/10.1002/pra2.510

Löffler, F., Wesp, V., König-Ries, B., & Klan, F. (2021). Dataset search in biodiversity research: Do metadata in data repositories reflect scholarly information needs? *PLOS ONE, 16*(3), e0246099. https://doi.org/10.1371/journal.pone.0246099

Lopez-Veyna, J. I., Sosa, V. J. S., & López-Arévalo, I. (2012). KESOSD: keyword search over structured data. *Proceedings of the Third International Workshop on Keyword Search on Structured Data, KEYS 2012, Scottsdale, AZ, USA, May 20, 2012*, 23–31. https://doi.org/10.1145/2254736. 2254743

Losee, R. M. (2006). Browsing mixed structured and unstructured data. *Information Processing & Management, 42*(2), 440–452. https://doi.org/10.1016/j.ipm.2005.02.001

Maddalena, E., Mizzaro, S., Scholer, F., & Turpin, A. (2017). On crowdsourcing relevance magnitudes for information retrieval evaluation. *ACM Transactions on Information Systems, 35*(3), 1–32. https://doi.org/10.1145/3002172

Mahyar, N., & Tory, M. (2014). Supporting communication and coordination in collaborative sensemaking. *IEEE Transactions on Visualization and Computer Graphics, 20*(12), 1633–1642. https:// doi.org/10.1109/TVCG.2014.2346573

Maitlis, S. (2005). The social processes of organizational sensemaking. *Academy of Management Journal, 48*(1), 21–49. https://doi.org/10.5465/amj.2005.15993111

Maitlis, S., & Christianson, M. (2014). Sensemaking in organizations: Taking stock and moving forward. *Academy of Management Annals, 8*(1), 57–125. https://doi.org/10.5465/19416520.2014. 873177

Malakis, S., & Kontogiannis, T. (2013). A sensemaking perspective on framing the mental picture of air traffic controllers. *Applied Ergonomics, 44*(2), 327–339. https://doi.org/10.1016/j.apergo. 2012.09.003

Marchionini, G. (1997). *Information seeking in electronic environments*. Cambridge university press. ISBN: 0521443725, 9780521443722

Marchionini, G. (2006). Exploratory search: From finding to understanding. *Commun. ACM, 49*(4), 41–46. https://doi.org/10.1145/1121949.1121979

Marchionini, G., Haas, S. W., Zhang, J., & Elsas, J. L. (2005). Accessing government statistical information. *IEEE Computer, 38*(12), 52–61. https://doi.org/10.1109/MC.2005.393

Marchionini, G., & Komlodi, A. (1998). Design of interfaces for information seeking. *Annual Review of Information Science and Technology, 33*, 89–130.

Marchionini, G., & White, R. (2007). Find what you need, understand what you find. *Int. J. Hum. Comput. Interaction, 23*(3), 205–237. https://doi.org/10.1080/10447310701702352

Marchionini, G., & White, R. W. (2009). Information-seeking support systems. *Computer, 42*(3), 30–32. https://doi.org/10.1109/MC.2009.88

Martín-Mora, E., Ellis, S., & Page, L. M. (2020). Use of web-based species occurrence information systems by academics and government professionals. *PloS One, 15*(7), e0236556. https://doi.org/10.1371/journal.pone.0236556

McCallum, S. H. (2006). *A Look at New Information Retrieval Protocols: SRU, OpenSearch/A9, CQL, and XQuery.* World Library and Information Congress: 72nd IFLA General Conference and Council.

McDermott, P. (2010). Building open government. *Government Information Quarterly, 27*(4), 401–413. https://doi.org/10.1016/j.giq.2010.07.002

Megler, V. M., & Maier, D. (2012). When big data leads to lost data. *Proceedings of the 5th Ph.D. Workshop on Information and Knowledge,* 1–8. https://doi.org/10.1145/2389686.2389688

Megler, V. M., Maier, D., Bottomly, D., White, L., McWeeney, S., & Wilmot, B. (2015). Data like this: Ranked search of genomic data vision paper. *2nd International Workshop on Exploratory Search in Databases and the Web, Explore DB 2015 - Proceedings.* https://doi.org/10.1145/2795218.2795221

Michener, W. K., Brunt, J. W., Helly, J. J., Kirchner, T. B., & Stafford, S. G. (1997). Nongeospatial Metadata for the Ecological Sciences. *Ecological Applications, 7*(1), 330–342. https://doi.org/10.1890/1051-0761(1997)007[0330:NMFTES]2.0.CO;2

Munzner, T. (2014). *Visualization Analysis and Design.* A K Peters/CRC Press. https://doi.org/10.1201/b17511

National Institutes of Health. (2003). *NIH data sharing policy and implementation guidance.* https://nihodoercomm.az1.qualtrics.com/jfe/form/SV_eypqaXlx2j1IY9T?Q_CHL=si&Q_CanScreenCapture=1

National Science Board. (2005). *Long-lived digital data dollections: Enabling research and education in the 21st century* (No. NSB0540). National Science Foundation. https://www.nsf.gov/pubs/2005/nsb0540/nsb0540.pdf

Neff, G., Tanweer, A., Fiore-Gartland, B., & Osburn, L. (2017). Critique and contribute: A practice-based framework for improving critical data studies and data science. *Big Data, 5*(2), 85–97. https://doi.org/10.1089/big.2016.0050

Nentidis, A., Bougiatiotis, K., Krithara, A., Paliouras, G., & Kakadiaris, I. (2017). Results of the fifth edition of the BioASQ Challenge. *BioNLP 2017,* 48–57. https://doi.org/10.18653/v1/W17-2306

Neumaier, S., Umbrich, J., & Polleres, A. (2016). Automated quality assessment of metadata across Open Data portals. *Journal of Data and Information Quality, 8*(1), 2:1–2:29. https://doi-org.uaccess.univie.ac.at/10.1145/2964909

Ninkov, A., Gregory, K., Peters, I., & Haustein, S. (2021). Datasets on DataCite - An initial bibliometric investigation. *Proceedings of the 18th International Conference of the International Society for Scientometrics and Informetrics.* International Conference on Scientometrics & Informatics (ISSI 2021), Leuven, Belgium (Virtual). https://zenodo.org/record/4730857

Noy, N., & Benjelloun, O. (2020). An analysis of online datasets using Dataset Search (Published, in Part, as a Dataset). *Google AI Blog.* http://ai.googleblog.com/2020/08/an-analysis-of-online-datasets-using.html

Odden, T. O. B., & Russ, R. S. (2019). Defining sensemaking: Bringing clarity to a fragmented theoretical construct. *Science Education, 103*(1), 187–205. https://doi.org/10.1002/sce.21452

Ohno-Machado, L., Sansone, S., Alter, G., Fore, I., Grethe, J., Xu, H., Gonzalez-beltran, A., Roccaserra, P., Gururaj, A. E., Bell, E., Soysal, E., Zong, N., & Kim, H. (2017). Finding useful data across multiple biomedical data repositories using DataMed. *Nature Genetics, 49*(6), 4–7. https://doi.org/10.1038/ng.3864

Oudshoorn, N. E.J. & Pinch, T. (2003). Introduction: How users and non-users matter. In Oud-shoorn, N.E.J. & Pinch, T. (Eds.), *How users matter. The co-construction of users and technology* (pp. 1–25). MIT Press. https://research.utwente.nl/en/publications/introduction-how-users-and-nonusers-matter(21c93d5a-f51b-47dd-a388-78583c2f8904).html

Page, L., Brin, S., Motwani, R., & Winograd, T. (1998). *The PageRank citation ranking: Bringing order to the web*. http://ilpubs.stanford.edu:8090/422/

Papenmeier, A., Krämer, T., Friedrich, T., Hienert, D., & Kern, D. (2021). Genuine information needs of social scientists looking for data. *Proceedings of the Association for Information Science and Technology, 58*(1), 292–302. https://doi.org/10.1002/pra2.457

Park, H., You, S., & Wolfram, D. (2018). Informal data citation for data sharing and reuse is more common than formal data citation in biomedical fields. *Journal of the Association for Information Science and Technology, 69*(11), 1346–1354. https://doi.org/10.1002/asi.24049

Park, T. K. (1993). The nature of relevance in information retrieval: An empirical study. *The Library Quarterly, 63*(3), 318–351. https://doi.org/10.1086/602592

Pasquetto, I. V., Borgman, C. L., & Wofford, M. F. (2019). Uses and reuses of scientific data: The data creators' advantage. *Harvard Data Science Review, 1*(2). https://doi.org/10.1162/99608f92.fc14bf2d

Pasquetto, I. V., Randles, B. M., & Borgman, C. L. (2017). On the reuse of scientific data. *Data Science Journal, 16*. https://doi.org/10.5334/dsj-2017-008

Paul, S. A., & Reddy, M. C. (2010). Understanding together: Sensemaking in collaborative information seeking. *Proceedings of the 2010 ACM Conference on Computer Supported Cooperative Work*, 321–330. https://doi.org/10.1145/1718918.1718976

Pepe, A., Goodman, A., Muench, A., Crosas, M., & Erdmann, C. (2014). How do astronomers share data? Reliability and persistence of datasets linked in AAS publications and a qualitative study of data practices among US astronomers. *PLoS One, 9*(8). https://doi.org/10.1371/journal.pone.0104798

Peters, I., Kraker, P., Lex, E., Gumpenberger, C., & Gorraiz, J. (2016). Research data explored: An extended analysis of citations and altmetrics. *Scientometrics, 107*(2), 723–744. https://doi.org/10.1007/s11192-016-1887-4

Pfister, H., & Blitzstein, J. (2015). Cs109/2015, Lectures 01-Introduction. In *GitHub repository*. GitHub. https://github.com/cs109/2015/tree/master/Lectures

Phillips, D., & Smit, M. (2021). Toward best practices for unstructured descriptions of research data. *Proceedings of the Association for Information Science and Technology, 58*(1), 303–314. https://doi.org/10.1002/pra2.458

Pienta, A., Akmon, D., Noble, J., Hoelter, L., & Jekielek, S. (2017). A data-driven approach to appraisal and selection at a domain data repository. *International Journal of Digital Curation, 12*(2), 362–375. https://doi.org/10.2218/ijdc.v12i2.500

Pirolli, P., & Card, S. (2005). The sensemaking process and leverage points for analyst technology as identified through cognitive task analysis. *Proceedings of International Conference on Intelligence Analysis, 5*, 2–4.

Pomerantz, J. (2015). *Metadata*. https://doi.org/10.7551/mitpress/10237.001.0001

Poth, C. N. (2019). Rigorous and ethical qualitative data reuse: Potential perils and promising practices. *International Journal of Qualitative Methods, 18*, 1609406919868870. https://doi.org/10.1177/1609406919868870

Ramdeen, S. (2017). Information seeking behavior of geologists when searching for physical samples. [University of North Carolina at Chapel Hill]. https://eric.ed.gov/?id=ED583308

Rezig, E. K., Vanterpool, A., Gadepally, V., Price, B., Cafarella, M., & Stonebraker, M. (2021). Towards data discovery by example. In V. Gadepally, T. Mattson, M. Stonebraker, T. Kraska, F. Wang, G. Luo, J. Kong, & A. Dubovitskaya (Eds.), *Heterogeneous Data Management, Polystores, and Analytics for Healthcare* (pp. 66–71). Springer International Publishing. https://doi.org/10. 1007/978-3-030-71055-2_6

Roberts, K., Gururaj, A. E., Chen, X., Pournejati, S., Hersh, W. R., Demner-Fushman, D., Ohno-Machado, L., Cohen, T., & Xu, H. (2017). Information retrieval for biomedical datasets: The 2016 bioCADDIE dataset retrieval challenge. *Database, 2017*(bax068). https://doi.org/10.1093/ database/bax068

Robinson-García, N., Jiménez-Contreras, E., & Torres-Salinas, D. (2016). Analyzing data citation practices using the data citation index. *Journal of the Association for Information Science and Technology, 67*(12), 2964–2975. https://doi.org/10.1002/asi.23529

Robinson-Garcia, N., Mongeon, P., Jeng, W., & Costas, R. (2017). DataCite as a novel bibliometric source: Coverage, strengths and limitations. *Journal of Informetrics, 11*(3), 841–854. https://doi. org/10.1016/j.joi.2017.07.003

Rowley, J. (2000). Product search in e-shopping: A review and research propositions. *Journal of Consumer Marketing, 17*(1), 20–35. https://doi.org/10.1108/07363760010309528

Russell, D. M. (2003). Learning to see, seeing to learn: Visual aspects of sensemaking. *Human Vision and Electronic Imaging VIII, Santa Clara, CA, USA, January 20, 2003,* 8–21. https://doi.org/10. 1117/12.501132

Russell, D. M. (2019). *The Joy of Search: A Google Insider's Guide to Going Beyond the Basics.* MIT Press. ISBN: 9780262546072

Russell, D. M., Stefik, M., Pirolli, P., & Card, S. K. (1993). The cost structure of sensemaking. *Human-Computer Interaction, INTERACT '93, IFIP TC13 International Conference on Human-Computer Interaction, 24–29 April 1993, Amsterdam, The Netherlands, Jointly Organised with ACM Conference on Human Aspects in Computing Systems CHI'93,* 269–276. https://doi.org/10. 1145/169059.169209

Sanderson, M., & Croft, W. B. (2012). The history of information retrieval research. *Proceedings of the IEEE, 100,* 1444–1451. https://doi.org/10.1109/JPROC.2012.2189916

Sands, A. E., Borgman, C. L., Wynholds, L., & Traweek, S. (2012). Follow the data: How astronomers use and reuse data. *Information, Interaction, Innovation: Celebrating the Past, Constructing the Present and Creating the Future - Proceedings of the 75th ASIS&T Annual Meeting, ASIST 2012, Baltimore, MD, USA, October 26–30, 2012, 49*(1), 1–3. https://doi.org/10.1002/ meet.14504901341

Sansone, S. A., Gonzalez-Beltran, A., Rocca-Serra, P., Alter, G., Grethe, J. S., Xu, H., Fore, I. M., Lyle, J., Gururaj, A. E., Chen, X., Kim, H. E., Zong, N., Li, Y., Liu, R., Ozyurt, I. B., & Ohno-Machado, L. (2017). DATS, the Data Tag Suite to enable discoverability of datasets. *Scientific Data, 4,* 1–8. https://doi.org/10.1038/sdata.2017.59

Sansone, S.-A., McQuilton, P., Rocca-Serra, P., Gonzalez-Beltran, A., Izzo, M., Lister, A. L., & Thurston, M. (2019). FAIRsharing as a community approach to standards, repositories and policies. *Nature Biotechnology, 37*(4), 358–367. https://doi.org/10.1038/s41587-019-0080-8

Saracevic, T. (1996). Modeling interaction in information retrieval (IR): A review and proposal. *Proceedings of the 59th Annual Meeting of the American Society for Information Science,* 3–9.

Saracevic, T. (1997). The stratified model of information retrieval interaction: Extension and applications. *Proceedings of the 60th Annual Meeting of the American Society for Information Science,* 313–327.

Savolainen, R. (1993). The sense-making theory: Reviewing the interests of a user-centered approach to information seeking and use. *Information Processing & Management, 29*(1), 13–28. https://doi.org/10.1016/0306-4573(93)90020-E

Savolainen, R. (2007). Information behavior and information practice: Reviewing the "umbrella concepts" of information-seeking studies. *The Library Quarterly, 77*(2), 109–132. https://doi.org/10. 1086/517840

Savolainen, R. (2018). Information-seeking processes as temporal developments: Comparison of stage-based and cyclic approaches. *Journal of the Association for Information Science and Technology.* https://doi.org/10.1002/jasist.24003

Scerri, A., Kuriakose, J., Deshmane, A. A., Stanger, M., Cotroneo, P., Moore, R., Raj Naik, & de Waard, A. (2016). Elsevier's approach to the bioCADDIE 2016 Dataset Retrieval Challenge. *Database,* 1–12. https://doi.org/10.1093/tropej/fmw080

Schmidt, B., Gemeinholzer, B., & Treloar, A. (2016). Open data in global environmental research: The Belmont Forum's open data survey. *PLoS One, 11*(1). https://doi.org/10.1371/journal.pone.0146695

Schoeffmann, K., Hudelist, M. A., & Huber, J. (2015). Video interaction tools: A survey of recent work. *ACM Computing Surveys, 48*(1), 14:1–14:34. https://doi.org/10.1145/2808796

Shneiderman, B. (1996). The eyes have it: A task by data type taxonomy for information visualizations. *Proceedings of the 1996 IEEE Symposium on Visual Languages, Boulder, Colorado, USA, September 3–6, 1996,* 336–343. https://doi.org/10.1109/VL.1996.545307

Silvertown, J. (2009). A new dawn for citizen science. *Trends in Ecology & Evolution, 24*(9), 467–471. https://doi.org/10.1016/j.tree.2009.03.017

Simpson, R., Page, K. R., & De Roure, D. (2014). Zooniverse: Observing the world's largest citizen science platform. *Proceedings of the 23rd International Conference on World Wide Web,* 1049–1054. https://doi.org/10.1145/2567948.2579215

Spolaôr, N., Lee, H. D., Takaki, W. S. R., Ensina, L. A., Coy, C. S. R., & Wu, F. C. (2020). A systematic review on content-based video retrieval. *Engineering Applications of Artificial Intelligence, 90,* 103557. https://doi.org/10.1016/j.engappai.2020.103557

Ragavan, S., Sarkar, A., & Gordon, A. D. (2021). Spreadsheet comprehension: Guesswork, giving up and going back to the author. *Proceedings of the 2021 CHI Conference on Human Factors in Computing Systems,* 1–21. https://doi.org/10.1145/3411764.3445634

Stasko, J. T., Görg, C., & Liu, Z. (2008). Jigsaw: Supporting investigative analysis through interactive visualization. *Information Visualization, 7*(2), 118–132. https://doi.org/10.1057/palgrave.ivs.9500180

Stock, W. G., & Stock, M. (2013). *Handbook of information science.* De Gruyter Saur.

Strecker, D. (2021). *Quantitative assessment of metadata collections of research data repositories* [Humboldt University]. https://edoc.hu-berlin.de/handle/18452/23590

Swan, A., & Brown, S. (2008). *To share or not to share: Publication and quality assurance of research data outputs* (pp. 56–56). http://www.rin.ac.uk/system/files/attachments/To-share-data-outputs-report.pdf

Talja, S., & Hansen, P. (2006). Information sharing. In A. Spink & C. Cole (Eds.), *New directions in human information behavior* (pp. 113–134). Springer Netherlands. https://doi.org/10.1007/1-4020-3670-1_7

Tang, R., Mehra, B., Du, J. T., & Zhao, Y. (Chris). (2019). Paradigm shift in information research. *Proceedings of the Association for Information Science and Technology, 56*(1), 578–581. https://doi.org/10.1002/pra2.96

Taylor, R. S. (1968). Question-negotiation and information seeking in libraries. *College & Research Libraries, 29*(3), 178–194.

Thelwall, M., & Kousha, K. (2016). Figshare: A universal repository for academic resource sharing? *Online Information Review, 40*(3), 333–346. https://doi.org/10.1108/OIR-06-2015-0190

Tombros, A., & Sanderson, M. (1998). Advantages of query biased summaries in information retrieval. *Proceedings of the 21st Annual International ACM SIGIR Conference on Research and Development in Information Retrieval - SIGIR '98,* 2–10. https://doi.org/10.1145/290941.290947

Toms, E. G. (2013). Task-based information searching and retrieval. In I. Ruthven & D. Kelly (Eds.), *Interactive Information Seeking, Behaviour and Retrieval* (1st ed., pp. 43–60). Facet. https://doi.org/10.29085/9781856049740.005

Tsagkias, M., King, T. H., Kallumadi, S., Murdock, V., & de Rijke, M. (2020). Challenges and research opportunities in eCommerce search and recommendations. *ACM SIGIR Forum, 54*(1), 1–23. https://doi.org/10.1145/3451964.3451966

Tukey, J. W. (1977). *Exploratory data analysis.* Addison-Wesley Pub. Co.

Vakkari, P. (2005). Task-based information searching. *Annual Review of Information Science and Technology, 37*(1), 413–464. https://doi.org/10.1002/aris.1440370110

Van House, N. (2004). Science and Technology Studies and Information Studies. *Annual Review of Information Science and Technology, 38,* 3–86.

van de Sandt, S., Dallmeier-Tiessen, S., Lavasa, A., & Petras, V. (2019). The definition of reuse. *Data Science Journal, 18,* 22. https://doi.org/10.5334/dsj-2019-022

Verhulst, S., & Young, A. (2016). Open data impact when demand and supply meet. In *Open Data Impact: When Demand and Supply Meets.* GOVLAB. https://odimpact.org/key-findings.html

Wallis, J. C., Borgman, C. L., Mayernik, M. S., Pepe, A., Ramanathan, N. c, & Hansen, M. (2007). Know thy sensor: Trust, data quality, and data integrity in scientific digital libraries. In L. Kovács, N. Fuhr, & C. Meghini (Eds.), *Research and Advanced Technology for Digital Libraries. ECDL 2007. Lecture Notes in Computer Science* (Vol. 4675, pp. 380–391). Springer, Berlin, Heidelberg. https://doi.org/10.1007/978-3-540-74851-9_32

Wallis, J. C., Rolando, E., & Borgman, C. L. (2013). If we share data, will anyone use them? Data sharing and reuse in the long tail of science and technology. *PloS ONE, 8*(7), e67332. https://doi.org/10.1371/journal.pone.0067332

Wang, S., & Zhang, K.-L. (2005). Searching databases with keywords. *Journal of Computer Science and Technology, 20*(1), 55–62. https://doi.org/10.1007/s11390-005-0006-4

White, R. W. (2016). *Interactions with search systems.* Cambridge University Press. ISBN: 1107034221

White, R. W., Jose, J. M., & Ruthven, I. (2003). A task-oriented study on the influencing effects of query-biased summarisation in web searching. *Information Processing & Management, 39*(5), 707–733. https://doi.org/10.1016/S0306-4573(02)00033-X

Wicherts, J. M., Borsboom, D., Kats, J., & Molenaar, D. (2006). The poor availability of psychological research data for reanalysis. *American Psychologist, 61*(7), 726–728. https://doi.org/10.1037/0003-066X.61.7.726

Wilkinson, M. D., Dumontier, M., Aalbersberg, Ij. J., Appleton, G., Axton, M., Baak, A., Blomberg, N., Boiten, J.-W., da Silva Santos, L. B., Bourne, P. E., & others. (2016). The FAIR Guiding Principles for scientific data management and stewardship. *Scientific Data, 3.* https://doi.org/10.1038/sdata.2016.18

Wilson, M. L., Kules, B., schraefel, m c, & Shneiderman, B. (2010). From keyword search to exploration: Designing future search interfaces for the Web. *Foundations and Trends in Web Science, 2*(1), 1–97. https://doi.org/10.1561/1800000003

Wilson, P. T. D. (1994). Information needs and uses: Fifty years of progress. In B. C. Vickery (Ed.), *Fifty years of information progress: A Journal of Documentation review* (pp. 15–51). Aslib.

Wilson, T. D. (1999). Models in information behaviour research. *Journal of Documentation, 55*(3), 249–270. https://doi.org/10.1108/EUM0000000007145

Wu, M. (2021). *Studies on eliciting data discovery context.* ASIS&T 84th Annual Meeting.

Wyatt, S. (2003). Non-users also matter: The construction of users and non-users of the Internet. In Oudshoorn, N. & Pinch, T. (Eds.), *How users matter: The co-construction of users and technology* (pp. 67–79). MIT Press. https://dare.uva.nl/personal/pure/en/publications/nonusers-also-matter-the-construction-of-users-and-nonusers-of-the-internet(9fc15327-9657-4e65-9af3-b5def673f4e9).html

Wynholds, L. A., Wallis, J. C., Borgman, C. L., Sands, A., & Traweek, S. (2012). Data, data use, and scientific inquiry: Two case studies of data practices. *Proceedings of the 12th ACM/IEEE-CS Joint Conference on Digital Libraries*, 19–22. https://doi.org/10.1145/2232817.2232822

Wynholds, L., Jr, D. S. F., Borgman, C. L., & Traweek, S. (2011). When use cases are not useful: Data practices, astronomy, and digital libraries. *Proceedings of the 2011 Joint International Conference on Digital Libraries, JCDL 2011, Ottawa, ON, Canada, June 13–17, 2011*, 383–386. https://doi.org/10.1145/1998076.1998146

Xiao, F., He, D., Chi, Y., Jeng, W., & Tomer, C. (2019). Challenges and supports for accessing open government datasets: Data guide for better open data access and uses. *Proceedings of the 2019 Conference on Human Information Interaction and Retrieval*, 313–317. https://doi.org/10.1145/3295750.3298958

Xiao, F., Ma, R., & He, D. (2020b). Task-based human-structured research data interaction: A discipline independent examination. *Proceedings of the Association for Information Science and Technology, 57*(1), e308. https://doi.org/10.1002/pra2.308

Xiao, F., Wang, Z., & He, D. (2020a). Understanding users' accessing behaviors to local Open Government Data via transaction log analysis. *Proceedings of the Association for Information Science and Technology, 57*(1), e278. https://doi.org/10.1002/pra2.278

Xie, I. (2008). *Interactive information retrieval in digital environments*. IGI Publishing. ISBN: 978-1-59904-242-8

Yalçin, M. A., Elmqvist, N., & Bederson, B. B. (2018). Keshif: Rapid and expressive tabular data exploration for novices. *IEEE Transactions on Visualization and Computer Graphics, 24*(8), 2339–2352. https://doi.org/10.1109/TVCG.2017.2723393

Yoon, A. (2014). Making a square fit into a circle: Researchers' experiences reusing qualitative data. *Proceedings of the American Society for Information Science and Technology, 51*(1), 1–4. https://doi.org/10.1002/meet.2014.14505101140

Zaveri, A., Rula, A., Maurino, A., Pietrobon, R., Lehmann, J., & Auer, S. (2016). Quality assessment for Linked Data: A Survey. *Semantic Web, 7*(1), 63–93. https://doi.org/10.3233/SW-150175

Zhang, S., & Balog, K. (2018). Ad hoc table retrieval using semantic similarity. *Proceedings of the 2018 World Wide Web Conference*, 1553–1562. https://doi.org/10.1145/3178876.3186067

Zhang, L., & Rui, Y. (2013). Image search—From thousands to billions in 20 years. *ACM Transactions on Multimedia Computing, Communications, and Applications, 9*(1s), 1–20. https://doi.org/10.1145/2490823

Zhang, M., Zhang, Y., Liu, L., Yu, L., Tsang, S., Tan, J., Yao, W., Kang, M. S., An, Y., & Fan, X. (2010). Gene Expression Browser: Large-scale and cross-experiment microarray data integration, management, search & visualization. *BMC Bioinformatics, 11*, 433–433. https://doi.org/10.1186/1471-2105-11-433

Zimmerman, A. (2007). Not by metadata alone: The use of diverse forms of knowledge to locate data for reuse. *Int. J. on Digital Libraries, 7*(1–2), 5–16. https://doi.org/10.1007/s00799-007-0015-8

Zimmerman, A. S. (2008). New knowledge from old data: The role of standards in the sharing and reuse of ecological data. *Science Technology and Human Values, 33*(5), 631–652. https://doi.org/10.1177/0162243907306704

Zuiderwijk, A., & Janssen, M. (2014). Open data policies, their implementation and impact: A framework for comparison. *Government Information Quarterly, 31*(1), 17–29. https://doi.org/10.1016/j.giq.2013.04.003

Zuiderwijk, A., Janssen, M., & Dwivedi, Y. K. (2015). Acceptance and use predictors of open data technologies: Drawing upon the unified theory of acceptance and use of technology. *Government Information Quarterly, 32*(4), 429–440. https://doi.org/10.1016/j.giq.2015.09.005

Printed in the United States
by Baker & Taylor Publisher Services